MW01503429

SAE
Electrical Standard
for Industrial Machinery
- Supplement to NFPA 79

SAE HS-1738

Published by:
Society of Automotive Engineers, Inc.
400 Commonwealth Drive
Warrendale, PA 15096-0001
U.S.A.
Phone: (412) 776-4841
Fax: (412) 776-5760

SAE - NFPA Machinery & Equipment Electrical Committee

Lynn F. Saunders	GM Worldwide Facilities.
Ray Butler	Sick Optics
John Cabrera	Ford Motor
Frank Dres	Vector Eng.
Mike Dwyer	Hoffman Eng.
Ron Fesl	Bussman
Dave Fisher	Allen Bradley
Pat Galuardi	Phoenix
Tony Keller	GM Delco Elect.
Ken Kramer	Ford Motor
Chuck Marino	Rittal Corp.
Mark Medici	Chrysler
Todd Mickley	Hoffman
Richard Mills	Sick Optics
Roberta Nelson	Data Instruments
John Orvis	Deere & Co.
Ken Paape	Westinghouse
Jim Peace	GE Co.
Terry Roberts	Daykin Elect.
Don Ruthman	Myron Zucker
Pete Schimmoeller	Pepperl & Fuchs
Tim Senter	JIC Electric
Fred Stevenson	Deere & Co.
Lynn Wallace	Eaton Corp.
Wayman Withrow	Cinncinnati Inc.
Jeff Zillisch	Wago Corp.
Darlene Crocker	SAE
Tom Northrup	

This SAE Supplement to the NFPA 79 - 1991 _Electrical Stand for Industrial Machinery_ has been developed to provide a document to be used in the design and application of electrical equipment in Automotive Industry applications.

In the past, organization like General Motors, Ford Motor, and Chrysler Corporation would develop complete electrical standards documents, rewriting the basic information contained in general industry standards.

The SAE Supplement document contains the basic industry standard document NFPA 79, and includes the additional Automotive requirements necessary.

The document is formatted in a manner that will increase the knowledge and understanding of the basic requirements while highlighting the reasons for the Automotive Industry supplements.

Part 1 of the SAE Supplement contains the NFPA 79-1991 text with the additions of several vertical bars in the margins to indicate revisions. The following legend indicates the type and are illustrated in the example below:

Legend:

Left Border Line (Thin)	NFPA 79 Text Revisions found in 1991 edition
Right Border Line (Thick)	SAE-Suggested Text Revisions for NFPA 79 1991 edition. Represents additions, deletions or replacements to existing text.
Insertion Bar, Right Border	SAE-Suggested Text Revisions for NFPA 79 1991 edition. Represents the addition of a new section, paragraph or addition to the end of existing paragraph.
Double Insertion Bar, Right Border	SAE-Suggested Text Revisions for NFPA 79 1991 edition. Represents the addition of more than one new section or paragraph.

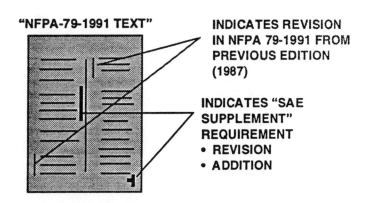

"NFPA-79-1991 TEXT"

INDICATES REVISION IN NFPA 79-1991 FROM PREVIOUS EDITION (1987)

INDICATES "SAE SUPPLEMENT" REQUIREMENT
• REVISION
• ADDITION

Part 2 of the document contains the SAE Supplement text. The left column contains the "requirement" information while the right column contains the "Rationale".

The rationale section of this Standard is intended to justify the additions, deletions, and changes to the NFPA 79 text. This rationale justifies why the automotive industry needs additional requirements over the NFPA 79 general industrial machinery standard. NFPA 79 is intended to be a safety standard and does not fully address issues of reliability, maintainability, and serviceability of automotive industrial machinery. It is not the intent of the rationale Section of this standard to be an application guide.

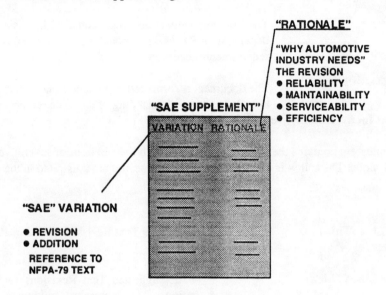

The Table of Contents for the NFPA 79 document follows. The revision shown for each section is the date of the SAE Supplement document developed in the process.

SAE SUPPLEMENT TO NFPA 79
ELECTRICAL STANDARDS FOR INDUSTRIAL MACHINERY

Contents

Preface

Chapter 1 General

Chapter 2 Diagrams, Instructions, and Nameplates

Chapter 3 General Operating Conditions

Chapter 4 Safeguarding of Personnel

Chapter 5 Supply Circuit Disconnecting Means

Chapter 6 Protection

Chapter 7 Control Circuits

Chapter 8 Control Equipment

Chapter 9 Control Enclosures and Compartments

Chapter 10 Location and Mounting of Control Equipment

Chapter 11 Operator's Control Stations and Equipment

Chapter 12 Accessories and Lighting

Chapter 13 Conductors

Chapter 14 Wiring Methods and Practices

Chapter 16 Motors and Motor Compartments

Chapter 17 Grounded Circuits and Equipment Grounding

Chapter 18 Electronic Equipment

Chapter 19 Referenced Publications

Appendix A Glossary of Terms

Appendix B Examples of Industrial Machinery Covered by NFPA 79

Appendix C Graphical Symbols

Appendix D Sample Electrical Diagrams

Appendix E Device and Component Designations

Appendix F Referenced Publications

Appendix G Equipment Data Form   *New in SAE*

Appendix H Application Guide for Power Factor Correction and   *New in SAE*
Harmonic Filtering

Response Form on SAE HSJ1738 Usage

Request for Revision of HSJ1738

NFPA 79
Electrical Standard for
Industrial Machinery
1991 Edition

Information on referenced publications can be found in Chapter 19 and Appendix F.

Preface

In September 1941, the metalworking machine tool industry wrote its first Electrical Standard to make machine tools safer to operate, more productive, less costly to maintain and to improve the quality and performance of their electrical components. That particular standard served as an American "War Standard".

To study the special electrical problems involved with machine tools, the Electrical Section of the National Fire Protection Association in 1941 sanctioned a Special Subcommittee on Wiring, Overcurrent Protection, and Control of Motor Operated Machine Tools. This Subcommittee, cooperating with machine tool builders, manufacturers of control equipment, and Underwriters' Laboratories, Inc., conducted tests and investigated the peculiar conditions involved with machine tools which might warrant exception to certain specific *National Electrical Code* requirements. This investigation resulted on August 4, 1942, in a Tentative Interim Amendment and first appeared in a 1943 Supplement to the 1940 Edition of the *National Electrical Code* as Article 670, "Machine Tools". It remained essentially unchanged through the 1959 edition.

Meanwhile, manufacturers of other types of industrial equipment erroneously began to follow the specialized practices permitted by Article 670. Late in 1952 a Technical Subcommittee on fundamentals of Electrically Operated Production Machinery and Material Handling and Processing Equipment for Fixed Locations was organized to attempt to group in one article the special requirements of this broad field. The extremely broad scope introduced so many problems that, in December 1956, the Technical Subcommittee was reorganized into an NFPA Committee whose scope was limited to Machine Tools and whose objective was the preparation of this NFPA standard with corresponding revisions in Article 670 in *the National Electrical Code*.

Modern machine tool electrical equipment may vary from that found on a single motor machine such as a drill press which performs a simple repetitive operation to that of the very large, multimotored automatic machines which involve highly complex electrical control systems, including electronic and solid-state devices and equipment. Generally these machines are especially designed, factory-wired and tested by the builder, and then erected in the plant in which they will be used. Because of their importance to the production of the plant, and their usual high cost, they are customarily provided with many safeguards and other devices, not often incorporated in the usual motor and control application as contemplated by the *National Electrical Code*.

Although these machines may be completely automatic, they are constantly attended, when operating, by a highly skilled operator. The machine usually incorporates many special devices to protect the operator, protect the machine and building against fires of electrical origin, protect the machine and the work in process against damage due to electrical failures, and protect against loss of production due to failure of a machine component. To provide these safeguards, it may be preferable to sacrifice deliberately a motor or some other component rather than to chance injury to the operator, the work, or the machine. It is because of such considerations that this standard varies from the basic concepts of motor protection as contained in the *National Electrical Code*.

As NFPA 79 evolved, it became apparent that certain classes of Light Industrial Machinery (i.e. small drill presses, bench grinders, sanders, etc.) were not appropriately covered. The NFPA 79-1977 standard recognized this problem and purposely excluded tools powered by two horsepower or less.

Subsequent to publication of the 1977 standard, a Light Industrial Machinery standard development activity was initiated by the Power Tool Institute. NFPA 79-1980 reflects this activity and appropriate requirements are now included in the standard.

In 1975, the Society of the Plastics Industry requested to have this standard enlarged in scope so as to include Plastics Machinery. A formal request was received by NFPA in September of 1978 and through combined efforts of NFPA 79 committees and representatives of the Society of Plastics Industry, the scope was broadened to include such machinery in the 1980 edition.

In June, 1981, the Joint Industrial Council Board of Directors acknowledged the dated state of the electrical and electronic standards and requested that NFPA 79 incorporate into its standard the material and topics covered by the JIC Electrical (EMP-1-67, EGP-1-67) and Electronic (EL-1-71) standards with the intention that the JIC standards eventually be declared superseded. The NFPA Standards Council approved the request with the stipulation that the material and topics incorporated from the JIC standards be limited to areas related to electrical shock and fire hazards. The 1985 edition reflects the incorporation of the appropriate material from the JIC Electrical (EMP-1-67, EGP-1-67) standards not previously covered. The 1991 edition includes additional references to international standards.

Chapter 1   General

1-1 Purpose.

(a) The purpose of this electrical standard is to provide detailed information for the application of electrical/electronic equipment, apparatus, or systems supplied as part of industrial machinery that will promote safety to life and property.

(b) This standard is not intended to limit or inhibit the advancement of the state of the art. Each type of machine has unique requirements which shall be accommodated to provide adequate safety.

1-2 Scope.

(a) The provisions of this standard shall apply to the electrical/electronic equipment, apparatus, or systems of industrial machine operating from a supply voltage of 600 volts or less, and commencing at the place of connection of the supply to the electrical equipment of the machines.

NOTE 1: In this standard, the term "electrical" includes both electrical and electronic equipment. Requirements that apply only to electronic equipment shall be so identified.

NOTE 2: The general terms "machine" and "machinery" as used throughout this standard mean industrial machinery.

(b) This standard shall not be considered adequate for machines intended for use in areas defined as hazardous (classified) locations by the NFPA 70-1990 *National Electrical Code*.

(c) This standard is not intended to apply to:

(1) Fixed or portable tools judged under the requirements of a testing laboratory acceptable to the authority having jurisdiction.
(2) Machines used in dwelling units.

(d) The installation of a machine and the wiring between component machines of "manufacturing systems/cells" shall be in accordance with NFPA 70, *National Electrical Code*, 1990 edition.

1-3 Definitions.

For purposes of this standard, the following definitions shall apply:

(a) A machine tool is defined as a power-driven not portable by hand, used to shape or form metal or plastic by cutting, impact, pressure, electrical techniques, or a combination of these processes.

Note: See Appendix B for examples of industrial machinery.

(b) Plastics machinery is defined as a power-driven machine not portable by hand, used to shape or form plastic by application of thermal energy, mechanical energy, or both, by cutting, impact, pressure, or a combination of these processes.

NOTE: See Appendix B for such types of plastics machinery.

(c) For the purposes of this standard, definitions of some other terms are given in Appendix A.

1-4 Other Standards. On any point for which specific provisions are not made in this standard, the provisions of the NFPA 70 *National Electrical Code,* 1990 edition, shall be observed. Other organizations having standards that may provide additional information are listed in Appendix F.

NOTE: NFPA 70E, *Standard for Electrical Safety Requirements for Employee Workplaces,* contains additional information.

Chapter 2   Diagrams, Instructions, and Nameplates

2-1 General.

(a) The following shall be included with the electrical drawings where required in Section 2-2.
    (1) Complete schematic diagram
    (2) Sequence of operations
    (3) Block diagram (where appropriate)
    (4) Equipment layout
    (5) Panel layouts
    (6) Interconnection diagram
    (7) Electronic schematics (where appropriate)
    (8) Parts list
    (9) Instruction and service manuals
    (10) Information for Safety Lockout Procedure (where appropriate).

(b) The following shall be furnished for reference where appropriate:
    (1) Lubrication diagram
    (2) Pneumatic diagram
    (3) Hydraulic diagram
    (4) Miscellaneous system diagrams (i.e. coolant, refrigerant, etc.)

(c) Where appropriate, a table of contents shall appear prominently on the first sheet and shall refer to all major sections of the electrical drawings.

2-2 Diagrams.

(a) Diagrams of the electrical system shall be provided. Any electrical symbols shall be in accordance with ANSI Y32.2. Any electrical symbols not shown in ANSI Y32.2 standards shall be separately shown and described on the diagrams. The symbols and identification of components and devices shall be consistent throughout all documents and on the machine.

Pertinent information such as motor horsepower, frame size, and speed shall be listed adjacent to its symbol.

NOTE: See Appendix D for examples of electrical diagrams.

(b) A cross-referencing scheme shall be used in conjunction with each relay, output device, pressure switch, etc. so that any contact associated with the device can be readily located on the diagrams.

(c) The functional description for each device shall be shown.

NOTE: See Appendix E for examples of devices and component designations.

(d) Switch symbols shall be shown on the electrical schematic diagrams with all utilities turned off (electric power, air, water, lubricant, etc.), the machine and its electrical equipment in its normal starting condition, and at 20 °C (68 °F) ambient. Control settings shall be shown on the drawings.

(e) Directly connected conductors shall be uniquely designated with the same alphanumeric reference. Conductors shall be identified in sequential order and in accordance with Chapter 14.

Exception No. 1: Conductors 18 AWG or less used in electronic assemblies need not be identified by an alphanumeric designation.

Exception No. 2: Where a plug is attached to a multiconductor cable, color coding shall be permitted to be substituted for the alphanumeric designation only at the plug end. Where color-coded multiconductor cable is used to wire identical components (e.g., limit switches), the color shall be consistent throughout. Where color coding is used, it shall be clearly indicated on the electrical diagrams.

(f) Circuits shall be shown in such a way as to facilitate the understanding of their function as well as maintenance and fault location. Control circuit devices shall be shown between vertical lines which represent control power wiring. The left vertical line shall be the control circuits common and the right line shall be the operating coils common, except where permitted by Chapter 7 design requirements. Control devices shall be shown on horizontal lines (rungs) between vertical lines. Parallel circuits shall be shown on separate horizontal lines directly adjacent (above or below) the original circuit.

(g) An interconnection diagram shall be provided on large systems having a number of separate enclosures or control stations. It shall provide full information about the external connections of all of the entire electrical equipment on the machine.

(h) Interlock wiring diagrams shall include devices, functions, and conductors in the circuit where it is used.

(i) Plug/receptacle pin identification shall be shown on the diagram(s).

2-3 Instructions.

(a) Information referring to the installation, sequence of operations, explanation of unique terms, list of recommended spare parts, maintenance instructions, and adjustment procedures of the machine's electrical equipment shall be furnished.

(b) The installation drawing(s) shall provide information necessary for preliminary machine and control setup. This includes information on supply cables, particularly if they are to be supplied by the end user; the size and purpose of any cable duct, raceway, or wireway that must be supplied by the end user; and the amount of space required to mount and maintain the machine and its electrical equipment.

(c) The description of the sequence of operations is required for electrical equipment comprising several interrelated functions. Where the machine can perform several sequences, the description of operation shall explain each of them and their interrelationship. Information shall be given that is necessary for the understanding of the electrical operation in conjunction with the mechanical, hydraulic, or pneumatic operation of the machine. Where the sequence of operations is programmed controlled, the information on programming the system for operation, maintenance, and repair shall be provided. A block diagram shall be permitted to be used to facilitate the understanding of the sequence of operations. The block diagram shows the electrical equipment together with its functional interrelationships by the use of symbols or blocks without necessarily showing all interconnections. References to the appropriate electrical diagram(s) shall be included on the block diagram.

(d) The parts list shall itemize recommended electrical spare parts together with the necessary data for ordering replacements.

(e) Maintenance and service instructions shall include:
    1. Information necessary for calibrating and adjusting components, devices and subassemblies.
    2. Operating instructions, including all information necessary to describe initial conditions and operations of the complete system.
    3. Troubleshooting information and suggestions for locating and replacing faulty components, suggested preventative maintenance schedules, and related data.

2-4 Markings. Nameplates, markings, and identification plates shall be of sufficient durability to withstand the environment involved.

2-5 Warning Marking.

(a) A warning marking shall be provided adjacent to the disconnect operating handle(s) where the disconnect(s) that is interlocked with the enclosure door does not deenergize all exposed parts when the disconnect(s) is in the "open (off)" position.

(b) When an attachment plug is used as the disconnecting means, a warning marking shall be attached to the control enclosure door or cover indicating that power shall be disconnected from the equipment before the enclosure is opened.

(c) Where the disconnecting means is remote from the control enclosure, a warning marking shall be attached to the enclosure door or cover indicating that the power shall be disconnected from the equipment before the enclosure is opened and that the enclosure is to be closed before the power is restored.

2-6 Machine Marking. The machine shall be marked with the builder's name, trademark, or other identification symbol.

2-7 Machine Nameplate Data.

(a) A permanent nameplate listing the machine serial number, supply voltage, phase, frequency, full load current, ampere rating of the largest motor or load, short-circuit interrupting capacity of the machine overcurrent protective device where furnished as part of the equipment, and the electrical diagram number(s) or the number of the index to the electrical diagrams (bill of material) shall be attached to the control equipment enclosure or machine where plainly visible after installation. Where more than one incoming supply circuit is to be provided, the nameplate shall state the above information for each circuit.

Exception: Where only a single motor or motor controller is used, the motor nameplate shall be permitted to serve as the electrical equipment nameplate where it is plainly visible.

(b) The full-load current shown on the nameplate shall not be less than the full-load currents for all motors and other equipment that can be in operation at the same time under normal conditions of use. Where unusual loads, duty cycles, etc. require oversized conductors, the required capacity shall be included in the full-load current specified on the nameplate.

(c) Where overcurrent protection is provided in accordance with Section 6-2, the machine shall be marked "overcurrent protection provided at machine supply terminals". A separate nameplate shall be permitted for this purpose.

2-8 Equipment Marking and Identification.

(a) Where equipment is removed from its original enclosure or is so placed that the manufacturer's identification plate is not easily read, an additional identification plate shall be attached to the machine or enclosure.

(b) Where a motor nameplate or connection diagram plate is not visible, additional identification shall be provided where it can be easily read.

(c) Nameplates, identification plates, or warning markings shall not be removed from the equipment.

(d) All control panel devices and components shall be plainly identified with the same designation as shown on the diagram(s). This identification shall be adjacent to (not on) the device or component.

Exception No. 1: Where the size or location of the devices makes individual identification impractical, group identification shall be used.

Exception No. 2: This section need not apply to machines on which the equipment consists only of a single motor, motor-controller, push-button station(s), and work light(s).

(e) All devices external to the control panel(s) shall be identified by a nameplate with the same designation as shown on the diagram(s) and mounted adjacent to (not on) the device.

Exception: Devices covered by Section 2-9.

(f) Terminations on multiconductor plugs and receptacles shall be plainly marked. The markings on the plugs/receptacles and on the drawings shall correspond.

(g) Where group protection as provided for in Section 6-5(d) is used, information specifying the short circuit protective device for each group protected motor branch circuit shall be included with the equipment.

2-9 Function Identification. Each control station device (push-button, indicating light, selector switch, etc.) shall be identified as to its function on or adjacent to the device.

NOTE: Consideration shall be given to the use of IEC symbols for push-buttons (see Appendix C for examples).

2-10 Equipment Grounding Terminal Marking. The equipment grounding terminal shall be identified with the word "GROUND", the letters "GND", "GRD", the letter "G", or by coloring the terminal GREEN.

NOTE: Some other standards recognize symbols C-2 (IEC 417-5019) and C-3 (IEC 417-5017).

Chapter 3   General Operating Conditions

3-1 General. This chapter describes the general requirements and conditions for the operation of the electrical equipment of the machine.

3-2 Electrical Components and Devices. Electrical components and devices shall be used or installed assuming the operating conditions of ambient temperature, altitude, humidity, and supply voltage outlined in this chapter, and within their design ratings, taking into account any de-rating stipulated by the component or device manufacturer

3-3 Ambient Operating Temperature. The electrical equipment shall be capable of operating in an ambient temperature range of 5 to 40 °C (41 to 104°F) under no load to full load conditions.

3-4 Altitude. The electrical equipment shall be suitable for operating correctly at altitudes up to 3300 feet (1000 m) above sea level.

3-5 Relative Humidity. The electrical equipment shall be capable of operating within a relative humidity range of 20-95 percent (non-condensing).

3-6 Transportation and Storage. The electrical equipment shall be designed to withstand storage and transportation temperatures within the range of -25 to +55°C (-13 to +131°F) and up to +65°C (+149 °F) for short periods of time not exceeding 24 hours. Suitable means shall be provided to prevent damage from excessive moisture, vibration, stress, and mechanical shock during shipment.

3-7 Supply Voltage. The electrical equipment shall operate satisfactorily at full-load as well as no-load under the following conditions:

(a) Voltage	90-100 percent of rated voltage
(b) Frequency	± 2 percent of rated frequency.
(c) Harmonic Distortion	Up to 10 percent of total RMS sum of the 2nd through 5th harmonics. Up to an additional 2 percent RMS sum of the 6th through 30th harmonic.
(d) Radio Frequency	2 percent RMS above 10KHZ.
(e) Impulse Voltage	200 percent peak voltage up to 1 ms duration with a rise time of 500 *ns* to 500 µs.
(f) Voltage Drop	Reductions of 50 percent of peak voltage for 1/2 cycle or 20 percent for 1 cycle. More than 1 sec between successive reductions.
(g) Micro-interruption	Supply disconnected or at zero voltage for 3 ms at any random time in the cycle. More than 1 sec between successive reductions.

3-8 Installation and Operating Conditions. The electrical equipment shall be installed and operated in accordance with the manufacturer's instructions. Any conditions that are outside the operating conditions specified in Chapter 3 shall be permitted where acceptable to both manufacturer and user.

Chapter 4   Safeguarding of Personnel

4-1 General. The electrical equipment shall provide safeguarding of persons against electrical shock both in normal service and in case of fault.

4-2 Safeguarding Against Electrical Shock in Normal Service.

(a) Live parts shall either be located inside enclosures as described in Chapters 9 and 10 or be completely covered by insulation that can only be removed by destruction (e.g. interconnecting cables).

(b) Enclosure interlocking as described in Sections 5-9 and 9-8 shall be provided.

(c) Grounding and bonding of the electrical equipment and machine members shall comply with Chapter 17 of this standard.

4-3   Safeguarding Against Electrical Shock by Machine Extra Low Voltage (MELV). Circuits of which not all live parts are protected against direct contact in normal service shall fulfill all of the following conditions:

(a) The highest voltage (with respect to ground) shall not exceed 30 VAC (RMS) or 30 VDC (with less than 10% ripple).

(b) The source of supply and all live parts and conductors of such circuits shall be separated or isolated from circuits with higher voltages by insulation rated for the maximum voltage used in the same part of the electrical equipment.

(c) One side of the circuit or one point of the source of supply of that circuit shall be connected to the grounding circuit associated with the higher voltages used on the machine and its related exposed conductive parts.

(d) Plugs and receptacles used in MELV circuits shall be chosen to preclude accidental connection to circuits having higher voltages.

4-4   Safeguarding Against Electrical Shock from Residual Voltages. Where the equipment includes elements which may retain residual voltages after being switched off, the voltage shall be reduced to below 50 volts within 1 minute after being disconnected.

4-5 Safeguarding against other hazards.

(a) Where hazards are identified for the specific machine to which the electrical equipment is present, provisions for the connection of the appropriate safeguards (e.g., guards and protective devices) shall be made. These safeguards shall function in accordance with the requirements of the specific industrial machine.

(b) Where the industrial machine is used in conjunction with other machines or equipment (e.g., in a manufacturing system/cell), provisions shall be made, where appropriate, to connect external emergency stop devices to the emergency stop circuit *(see section 7-6)*. Where appropriate, provisions shall also be made to indicate the status of the emergency stop circuit to other machines and associated equipment.

Chapter 5   Supply Circuit Disconnecting Means

5-1 General Requirements. A disconnecting means shall be provided for each incoming supply circuit.

5-2 Type.

(a) The disconnecting means shall be manually operable and shall be a fusible or nonfusible motor circuit switch or a circuit breaker in accordance with Sections 5-3 through 5-10.

(b) An attachment plug in accordance with Section 5-11.

5-3 Rating.

(a) The ampacity of the disconnecting means shall not be less than 115 percent of the sum of the full-load currents required for all equipment which may be in operation at the same time under normal conditions of use.

(b) The interrupting capacity of the disconnecting means shall not be less than the sum of the locked-rotor current of the largest motor plus the full-load current of all other connected operating equipment.

(c) Fusible motor-circuit switches or circuit breakers shall be applied in accordance with Chapter 6 of this standard.

5-4 Position Indication. The disconnecting means shall plainly indicate whether it is in the open (off) or closed position.

5-5 Supply Conductors to be Disconnected. Each disconnecting means shall disconnect all ungrounded conductors of a single supply circuit simultaneously. Where there is more than one source, additional individual disconnecting means shall be provided for each supply circuit, so that all supply to the machine may be interrupted.

5-6 Connections to Supply Lines. Incoming supply line conductors shall terminate at the disconnecting means with no connection to terminal blocks or other devices ahead of the disconnecting means.

5-7 Exposed Live Parts. There shall be no exposed live parts with the disconnecting means in the open (off) position.

NOTE: See Exception to Section 7-1 .

5-8 Mounting.

(a) The disconnecting means shall be mounted within the control enclosure or adjacent thereto. Where mounted within the control enclosure, the disconnecting means shall be mounted at the top of the control panel with no other equipment mounted directly above it. Wire bending space shall be provided as required by NFPA 70, *National Electrical Code.* Section 430-10(b). Space shall be determined by maximum wire size of incoming lines or by maximum capacity of line lugs on the disconnecting means.

Exception No. 1: In plastics extrusion machinery (extruders, film casting machines, film and sheet winding equipment, wire coating machinery, and sheet line and pull rope equipment ONLY-see Appendix B-2) where the design configuration of the enclosure may preclude mounting the disconnect as the uppermost component:
a. Live parts shall be guarded against accidental contact.
b. Barriers shall be placed in all enclosures to isolate the supply circuit conductors and terminals from other internal conductors and components.

Exception No. 2: Machines with a motor(s) totaling two horsepower or less shall be permitted to be connected to a remotely mounted disconnecting means through a flexible cord, cable, or conduit provided the disconnecting means is within sight of, readily accessible to, and no more than 20 feet (6 m) from the machine operator.

(b) Where two or more disconnecting means are provided within the control enclosure for multiple supply circuits, they shall be grouped in one location.

5-9 Interlocking.

(a) Each disconnecting means shall be mechanically or electrically interlocked, or both, with the control enclosure doors. Interlocking shall be reactivated automatically when panel doors are closed.

Exception No 1: A disconnecting means used only for maintenance lighting circuits within control enclosures shall not be required to be interlocked with the control enclosure. The marking requirements of Section 2-5(a) shall apply.

Exception No. 2: Where an attachment plug is used as the disconnecting means in accordance with Section 5-11.

Exception No. 3: A disconnecting means used for power supply circuits within control enclosures to memory elements and their support logic requiring power at all times to maintain the storage of information shall not be required to be interlocked with the control enclosure doors. The marking requirements of Section 2-4(a) shall apply.

Exception No. 4: Where a remotely mounted disconnecting means is permitted as per Section 5-8, Exception No. 2, interlocking shall not be required provided that a tool is required to open the enclosure door and a label is attached to the control enclosure warning of dangerous voltage inside and advising disconnection of the power before opening.

Interlocking shall be provided between the disconnecting means and its associated door to accomplish both of the following:
(1) To prevent closing of the disconnecting means while the enclosure door is open, unless an interlock is operated by deliberate action
(2) To prevent closing of the disconnecting means while the door is in the initial latch position or until the door hardware is fully engaged.

All doors on multiple-door enclosures shall be interlocked simultaneously with the door that is interlocked with the main disconnecting means.

(b) Where there are two or more sources of power to the equipment or where there are two or more independent disconnecting means, power wiring from each disconnecting means shall be run in separate conduit and shall not terminate in or pass through common junction boxes.

5-10 Operating Handle.

(a) The operating handle of the disconnecting means shall be readily accessible.

(b) The center of the grip of the operating handle of the disconnecting means, when it is in its highest position,

shall be not more than 6 1/2 feet (2 m) above the floor. A permanent operating platform, readily accessible by means of a permanent stair or ladder, shall be considered as the floor for the purpose of this requirement.

(c) The operating handle shall be capable of being locked only in the open (off) position.

(d) When the control enclosure door is closed, the operating handle shall positively indicate whether the disconnecting means is in the open (off) or closed position.

5-11 Attachment Plug and Receptacle. An attachment plug and receptacle shall be permitted as a disconnecting means providing all of the following conditions are complied with:

(a) The motor(s) on the machine shall total two horsepower or less.

(b) The supply voltage shall not exceed 150 volts to ground.

(c) DC shall not be used.

(d) The ampacity of the attachment plug shall not be less than 115 percent of the sum of the full-load currents required for all equipment which may be in operation at the same time under normal conditions of use.

(e) The attachment plug shall be single voltage rated.

(f) The attachment plug shall be provided with a grounding pole and so constructed that the grounding pole is made before any current-carrying poles are made and is not broken until all current-carrying poles of the attachment plug have been disconnected. A grounding pole shall not be used as a current-carrying part.

(g) The attachment plug shall be in sight of the operator's station and readily accessible.

(h) The marking requirements of Section 2-5 (b) shall apply.

Chapter 6   Protection

6-1 Machine Circuits. Diagram 6-1 shows typical circuits acceptable for protection of machine motors, resistive heating loads, and controls. Protective interlocks are not shown.

6-2 Supply Conductor and Machine Overcurrent Protection. The overcurrent protection as shown in line C of Diagram 6-1, Figures I through IV inclusive, may or may not be furnished as part of the machine. Where furnished as part of the machine, it shall consist of a single circuit breaker or set of fuses, and the machine shall bear the marking required in Section 2-7(c).

6-3 Additional Overcurrent Protection. Where required, the additional overcurrent protection shown in Line D of Diagram 6-1, Figures III and IV, shall be provided as part of the machine control.
Such overcurrent protection (fuse or overcurrent trip unit of a circuit breaker) shall be placed in each ungrounded branch circuit conductor. A circuit breaker shall open all ungrounded conductors of the branch circuit.

6-4 Location of Protective Devices. Overcurrent protective devices shall be located at the point where the conductor to be protected receives its supply.

Exception No. 1: Where all of the following conditions are complied with:

(1) the conductor has an ampacity of at least one-third (1/3) that of the conductor from which it is supplied

(2) the conductor is suitably protected from physical damage

(3) the conductor is not over 25 feet (7.6 m) long

(4) the conductor terminates in a single circuit breaker or set of fuses

Exception No. 2: Where all of the following conditions are complied with:

(1) the conductor has an ampacity of not less than the sum of the maximum continuous load currents supplied

(2) the conductor is not over 10 feet (3 m) long

(3) the conductor does not extend beyond the control panel enclosure

(4) the conductor terminates in a single circuit breaker or set of fuses.

6-5 Motor Branch Circuits.

(a) The overcurrent protective device for a branch circuit supplying a single motor shall be capable of carrying the starting current of the motor. Overcurrent protection shall be considered as being obtained when the overcurrent device has a rating or setting not exceeding the values given in Table 6-5(a). Where the overcurrent protection specified in the table is not sufficient for the starting current of the motor, it shall be permitted to be increased to a maximum of 400 percent of the motor full-load current for inverse time circuit breakers and non-time delay fuses, and a maximum of 225 percent for time delay or dual element fuses, and a maximum of 1300 percent for instantaneous trip breakers.

Exception: Where the values for the branch-circuit short-circuit and ground-fault protective devices determined by Table 6-5(a) do not correspond to the standard sizes or rating of fuses, non-adjustable circuit breakers or thermal protective devices, or possible settings of adjustable circuit breakers adequate to carry the load, the next higher size, rating, or setting shall be permitted.

TYPICAL DIAGRAMS—CONSULT TEXT

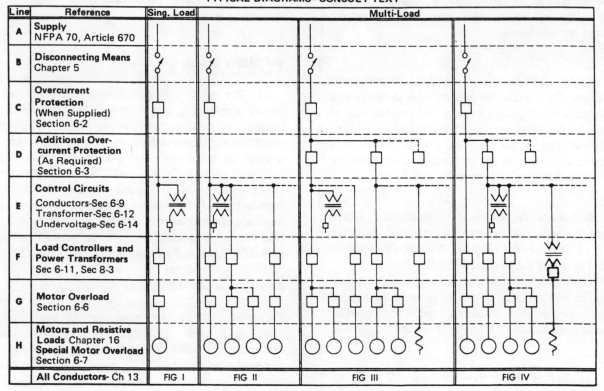

Diagram 6-1   Protection of Machine Electrical Circuits.

Table 6-5(a) Maximum Rating or Setting of Motor Branch-Circuit Short-Circuit and Ground-Fault Protective Devices

Type of Motor	Non-time Delay Fuse	Dual Element (Time Delay) Fuse	Instantaneous Trip Breaker	Inverse Time Breaker
Single-phase, all types				
No code letter	300	175	700	250
All ac single-phase and polyphase squirrel-cage and synchronous motors with full-voltage, resistor or reactor starting:				
No code letter	300	175	700	250
Code letter F to V	300	175	700	250
Code letter B to E	250	175	700	200
Code letter A	150	150	700	150
All ac squirrel-cage and synchronous motors with autotransformer starting:				
Not more than 30 amps				
No code letter	250	175	700	200
More than 30 amps				
No code letter	200	175	700	200
Code letter F to V	250	175	700	200
Code letter B to E	200	175	700	200
Code letter A	150	150	700	150
High-reactance squirrel-cage				
Not more than 30 amps				
No code letter	250	175	700	250
More than 30 amps				
No code letter	200	175	700	200
Wound motor-				
No code letter	150	150	700	150
Direct current (constant voltage)				
Not more than 50 hp				
No code letter	150	150	250	150
More than 50 hp				
No code letter	150	150	175	150

Percent of Full-Load Current

NOTE: Rating or Setting for Individual Motor Circuit. The motor branch-circuit short-circuit and ground-fault protective device shall be capable of carrying the starting current of the motor. The required protection shall be considered as being obtained where the protective device has a ratting or setting not exceeding the values given in the above table.

An instantaneous trip circuit breaker shall be used only if adjustable, if part of a combination controller having motor-running overload and also short-circuit and ground-fault protection in each conductor, and if the combination is especially identified.

(b) Several motors each not exceeding 1 horsepower in rating shall be permitted on a nominal 120-volt branch circuit protected at not over 20 amperes or a branch circuit of 600 volts, nominal, or less, protected at not over 15 amperes, where all of the following conditions are met:

(1) The full-load rating of each motor does not exceed 60 amperes.

(2) The rating of the branch-circuit short-circuit and ground-fault protective device marked on any of the controllers is not exceeded.

(3) Individual overload protection conforms to Section 6-6.

(c) Where the branch-circuit and ground-fault protective device is selected not to exceed that allowed in Section 6-5 (a) for the motor of the smallest rating, two or more motors or one or more motors and other load(s), with each motor having individual overload protection, shall be permitted to be connected to a branch circuit where it can be determined that the branch circuit short-circuit and ground-fault protective device will not open under the most severe normal conditions of service that might be encountered.

(d) Two or more motors and their control equipment shall be permitted to be connected to a single branch circuit where short-circuit and ground-fault protection is provided by a single inverse-time circuit breaker or a single set of fuses, provided both of the following conditions are met:

(1) Each motor controller and overload device is listed for group installation with specified short-circuit withstand ratings.

NOTE: The short-circuit withstand rating includes:

(a) The class and rating of the short-circuit protective device.

(b) The maximum nominal application voltage.

(c) The maximum available fault current.

(2) The rating or setting of the overcurrent device does not exceed the values in Table 6-5(d) for the smallest conductor in the circuit.

Table 6-5(d) Relationship between Conductor Size and Maximum Rating or Short Circuit Protective Device for Power Circuits

Conductor Size AWG	Max Rating Non-Time Delay Fuse or Inverse Time Circuit Breaker	Time Delay or Dual Element Fuse
14	60	30
12	80	40
10	100	50
8	150	80
6	200	100
4	250	125
3	300	150
2	350	175
1	400	200
0	500	250
2/0	600	300
3/0	700	350
4/0	800	400

6-6 Motor Overload.

(a) Overload devices shall be provided to protect each motor, motor controller, and branch-circuit conductor against excessive heating due to motor overloads or failure to start.

(b) Resetting of the overload device shall not restart the motor.

Exception: Where there is only a single motor of two horsepower or less on the machine, an overload reset operator mounted on the motor shall be permitted to restart the motor provided that the distance between the overload reset operator and the machine start pushbutton operator is 12 inches (300 mm) or less and a suitable warning label is attached on or adjacent to the overload reset operator.

(c) The minimum number and location of running overcurrent units shall be determined from Table 6-6(c).

Table 6-6(c) Running Overcurrent Units

Kind of Motor	Supply System	Number and Location of Overcurrent Units (such as trip coils relays, or thermal cutouts)
1-phase ac or dc	2-wire, 1-phase ac or dc ungrounded	1 in either conductor
1-phase ac or dc	2-wire, 1-phase ac or dc, one conductor grounded	1 in ungrounded conductor
1-phase ac or dc	3-wire, 1-phase ac or dc, grounded-neutral	1 in either ungrounded conductor
3-phase ac	Any 3-phase	*3, one in each phase

*Exception: *Unless protected by other approved means*

NOTE: For 2-phase power supply systems see the *NEC*, Section 430-37.

6-7 Motor Overload, Special Duty. Short-time rated motors or high-reversing duty motors that cannot be adequately protected by external overload devices shall be protected by a thermal device mounted in the motor and sensitive to the temperature of the motor, or to both motor temperature and current.

Motors that are an integral part of a refrigeration compressor of the hermetic or semihermetic type shall be protected per the compressor manufacturer's recommendations.

6-8 Resistance Heating Branch Circuits.

(a) If the branch-circuit supplies a single nonmotor-operated load rated at 16.7 amperes or more, the overcurrent device rating shall not exceed 150 percent of the load rating.

(b) Electric machines employing resistance-type heating elements rated at more than 48 amperes shall have the heating elements subdivided. Each subdivided load shall not exceed 48 amperes and shall be protected at no more than 60 amperes.

Exception: A single sheath-type heating element requiring more than 48 amperes shall be protected at not more than 125 percent of the load where the element is integral with and enclosed within the machine housing.

(c) The supplementary overcurrent protective devices shall be: (1) installed within or on the machinery or provided as a separate assembly; and (2) accessible but need not be readily accessible; and (3) suitable for branch-circuit protection.

(d) The main conductors supplying these overcurrent protective devices shall be considered branch-circuit conductors.

6-9 Control Circuit Conductors.

(a) *General.* A control circuit tapped from the load side of a branch-circuit short-circuit and ground-fault protective device(s) and functioning to control the load(s) connected to that branch circuit shall be protected against overcurrent in accordance with this section. Such a tapped control circuit shall not be considered to be a branch circuit and shall be permitted to be protected by either a supplementary or branch-circuit overcurrent protective device(s).

(b) Conductor *Protection.*

(1) Conductors larger than No. 14 shall be protected against overcurrent in accordance with their ampacities. The ampacities for control circuit conductors No. 14 and larger shall be those given in Table 13-5(a).

(2) Conductors of Nos. 18, 16, and 14 shall be considered as protected by an overcurrent device(s) of not more than 20 amperes of rating.

Exception No. 1 for (1) and (2) above: Conductors that do not extend beyond the enclosure shall be considered protected by the load branch-circuit short-circuit and ground-fault protective device(s) where the rating of the protective device(s) is not more than 400 percent of the ampacity of the control circuit conductor for conductors No. 14 and larger, or not more than 25 amperes for No. 18 and 40 amperes for No. 16. The ampacities for conductors No. 14 and larger shall be the values given in Table 13-5(a).

Exception No. 2 for (1) and (2) above: Conductors of No. 14 and larger that extend beyond the enclosure shall be considered protected by the load branch-circuit short-circuit and ground-fault protective device(s) where the rating of the protective device(s) is not more than 300 percent of the ampacity of the control circuit conductors. The ampacities shall be the values given in Table 13-5(a).

Exception No. 3 for (1) and (2) above: Conductors supplied by the secondary side of a single-phase transformer having a 2-wire (single voltage) secondary shall be considered protected by overcurrent protection provided on the primary (supply) side of the transformer, provided this protection is in accordance with Section 6-12 and does not exceed the value determined by multiplying the secondary conductor ampacity by the secondary-to-primary voltage ratio. Transformer secondary conductors (other than 2-wire) are not considered to be protected by the primary overcurrent protection.

Exception No. 4 for (1) and (2) above: Conductors of control circuits shall be considered as protected by the motor branch-circuit short-circuit and ground-fault protective device(s) where the opening of the control circuit would create a hazard, as for example, the control circuit of a magnetic chuck and the like.

6-10 Lighting Branch Circuits. Overcurrent protection for lighting branch-circuits shall not exceed 15 amperes.

6-11 Power Transformer. As used in this section, the word "transformer" shall mean a power transformer or polyphase bank of two or three single-phase power transformers operating as a unit to supply power to loads other than control circuit devices.

(a) *Primary.* Each 600 volt or less transformer shall be protected by an individual overcurrent device on the primary side rated or set at no more than 125 percent of the rated primary current of the transformer.

Exception No. 1: Where the rated primary current of a transformer is 9 amperes or more and 125 percent of this current does not correspond to a standard rating of a fuse or nonadjustable circuit breaker, the next higher standard rating shall be permitted. Where the rated primary current is less than 9 amperes, an overcurrent device rated or set at not more than 167 percent of the primary current shall be permitted.

Where the rated primary current is less than 2 amperes, an overcurrent device rated or set at not more than 300 percent shall be permitted.

Exception No. 2: An individual overcurrent device shall not be required where the primary circuit overcurrent device provides the protection specified in this section.

Exception No. 3: As provided in (b) below.

(b) *Primary and Secondary.* A transformer, 600 volts or less, having an overcurrent device on the secondary side rated or set at not more than 125 percent of the rated secondary current of the transformer shall not be required to have an individual overcurrent device on the primary side if the primary feeder overcurrent device is rated or set at a current value not more than 250 percent of the rated primary current of the transformer.

A transformer, 600 volts or less, equipped with coordinated thermal overload protection by the manufacturer and arranged to interrupt the primary current shall not be required to have an individual overcurrent device on the primary side if the primary feeder overcurrent device is rated or set at a current value not more than 6 times the rated current of the transformer for transformers having not more than 6 percent impedance, and not more than 4 times the rated current of the transformer for transformers having more than 6 but not more than 10 percent impedance.

Exception: Where the rated secondary current of a transformer is 9 amperes or more and 125 percent of this current does not correspond to a standard rating of a fuse or nonadjustable circuit breaker, the next highest standard rating shall be permitted.

Where the rated secondary current is less than 9 amperes, an overcurrent device rated or set at not more than 167 percent of the rated secondary current shall be permitted.

6-12 Control Circuit Transformer.

(a) Where a control circuit transformer is provided, the transformer shall be protected in accordance with Table 6-12.

Exception No. 1: Where the control circuit transformer is an integral part of the motor controller and is located within the motor controller enclosure, and where an overcurrent device(s) rated or set at not more than 200 percent of the rated secondary current of the transformer is provided in the secondary circuit.

Exception No. 2: Where the transformer supplies a Class 1 power-limited, Class 2 or Class 3 remote-control circuit.

Exception No. 3: Overcurrent protection shall be omitted where the opening of the control circuit would create a hazard, *as for example, the control circuit of a magnetic chuck and the like.*

(b) Where the circuit is grounded, the protective device(s) shall be located only in the ungrounded side.

(c) Where multiple overcurrent protective devices are used to protect individual branch circuits, and the sum of the current ratings of these overload protective devices exceeds the current allowed in Table 6-12, a single overload protective device complying with Table 6-12 shall be placed in the circuit ahead of the multiple protective devices. The rating or setting of the overcurrent protective device shall not exceed the values in Table 6-12 for the rating of the control transformer.

(d) Control circuit voltage derived from a power transformer shall be permitted.

Table 6-12 Control Transformer Overcurrent Protection (120 Volt Secondary)

Control Transformer Size, Volt-Amperes	Maximum Rating, Amperes
50	0.5
100	1.0
150	1.6
200	2.0
250	2.5
300	3.2
500	5
750	8
1000	10
1250	12
1500	15
2000	20
3000	30
5000	50

NOTE: For transformers larger than 5000 volt-amperes, the protective device rating shall be based on 125 percent of the secondary current rating of the transformer.

6-13 Common Overcurrent Device. The use of the same overcurrent device to provide the protection called for in Sections 6-9, 6-10, 6-11, and 6-12 shall be permitted.

6-14 Undervoltage Protection.

(a) In cases where a voltage drop below a specified level can cause malfunctioning of the electrical equipment, a minimum voltage device or detector that ensures appropriate protection at a predetermined voltage level shall be provided.

(b) The electrical equipment shall be designed to prevent automatic restart of any machine motion or cycles after power has been restored to required operating levels.

Exception No. 1: Blower motors where moving parts are fully guarded.

Exception No. 2: Coolant pumps.

Exception No. 3: Pumps utilized to maintain the raw materials in a workable condition.

Exception No 4: Compressor pumpdown circuits

(c) In an unsupported extrusion system such as blown film, sheet, or pipe, and where the operation of the machine can allow for an interruption of the voltage during a fraction of a second, a delayed no-voltage device shall be permitted. The delayed interruption and the reclosing shall in no way hinder instantaneous interruption by the control and operating devices (limit switches, relays, pushbuttons, etc.).

6-15 Adjustable Speed Drive System. The incoming branch circuit or feeder to power conversion equipment included as part of an adjustable speed drive system shall be based on the rated input to the power conversion equipment. Where the power conversion equipment provides overload protection for the motor, additional overload protection is not required.

6-16 Motor Overspeed Protection. Unless the inherent characteristics of the motor, or the controller or both are such as to limit the speed adequately, drive systems motors shall include protection against motor overspeed.

Overspeed protection means include, but are not necessarily limited to. the following:

(a) A mechanical overspeed device incorporated in the drive to remove armature voltage on motor overspeed.

(b) An electrical overspeed detector that will remove armature voltage on motor overspeed.

(c) Field loss detection to remove armature voltage upon the loss of field current.

(d) Voltage-limiting speed-regulated drives that operate with constant full field. In this case, protection is obtained individually for the loss of field or tachometer feedback; however, protection against simultaneous loss of field and tachometer is not provided.

The safe operating speed of the driven equipment may be lees than that of the motor. In this case, the user should coordinate with the drive manufacturer to obtain the most suitable means of limiting operation to safe operating speed.

6-17 Equipment Overspeed Protection. Where the safe operating speed of the equipment is less than that of the drive motor, means shall be provided to limit the speed of the equipment.

Chapter 7   Control Circuits

7-1 Source of Control Power. The source of supply for all control circuits shall be taken from the load side of the main disconnecting means.

Exception: Power supply to memory elements and their support logic requiring power at all times to maintain the storage of information shall be permitted to be taken from the line side of the main disconnecting means or other power source. The marking requirements of Section 2-5(a) shall apply.

7-2 Control Circuit Voltages.

(a) Alternating-current (ac) control voltage shall be 120 volts or less, single phase, obtained from a transformer with an isolated secondary winding.

Exception No. 1: Other voltages shall be permitted, where necessary, for the operation of electronic, precision, static, or similar devices used in the control circuit.

Exception No. 2: Exposed, grounded control circuits shall be permitted when supplied by a transformer having a primary rating of not more than 120 volts, a secondary rating of not more than 25 volts, and a capacity of not more than 50 volt-amperes.

Exception No. 3: Any electro-mechanical magnetic device having an inrush current exceeding 20 amperes at 120 volts shall be permitted to be energized at line voltage through contactor or relay contacts. The contactor or relay contacts shall break both sides of the line voltage circuit to the magnetic device. The relay coil shall be connected to the control circuit.

(b) Direct-current (dc) control voltage shall be 250 volts or less.

Exception: Other voltages shall be permitted, where necessary, for the operation of electronic, precision, static, or similar devices used in the control circuit.

7-3 Grounding of Control Circuits. Grounded or ungrounded control circuits shall be permitted as provided in Section 17-7. Ground faults on any control circuit shall not cause unintentional starting or dangerous movements, or prevent stopping of the machine.

7-4 Connection of Control Devices.

(a) All operating coils of electromechanical magnetic devices and indicator lamps (or transformer primary windings for indicator lamps) shall be directly connected between the coil and the other side of the control circuit. All control circuit contacts shall be connected between the coil and the other side of the control circuit.

Exception No. 1: Electrical interlock contacts on multispeed motor controllers where the wiring to these contacts does not extend beyond the control enclosure.

Exception No. 2: Overload relay contacts where the wiring to these contacts does not extend beyond the control enclosure.

Exception No. 3: Contacts of multipole control circuit switching devices that simultaneously open both sides of the control circuit.

Exception No. 4: Ground test switching device contacts in ungrounded control circuits.

Exception No. 5: Solenoid test switching device contacts in ungrounded circuits.

Exception No. 6: Coils or contacts used in electronic control circuits where the wiring to these coils or contacts does not extend beyond the control enclosure.

Exception No. 7: "Run" pushbuttons for two-hand operation, such as for presses having ground detection circuits and overcurrent protection in each conductor.

(b) Contacts shall not be connected in parallel to increase ampacity.

7-5 Stop Circuits.

(a) Stop functions shall be initiated through de-energization rather than energization of control devices.

(b) Stop functions shall override their related start functions.

(c) Each machine shall incorporate an emergency stop circuit with at least one manually operated emergency stop device. *(See Section 7-6 for requirements of the emergency stop circuit.)*

Exception: Where the emergency stop function duplicates the normal or zone (where provided) stop function, a separate emergency stop function shall not be required.

(d) A zone stop is an optional stop function that may be requested for applications of robots or similar equipment, where protection is required, that may be used where safeguards and human detection devices are required. *(See Section 7-7 for requirements of the zone stop circuit.)*

NOTE: In IEC 204-1, this stop function is a safety stop.

(e) A normal stop circuit(s) shall be provided to stop motion under normal operating conditions. [*See Section 7-8 for requirements of the normal stop circuit(s).*]

7-6 Emergency Stop.

(a) The emergency stop circuit shall function by removing power to actuators which cause hazardous conditions as quickly as possible without creating other hazards (i.e. by providing means requiring no external power).

(b) The functioning of any braking system fitted to the machine to stop it more rapidly shall not be prevented by actuating the emergency stop circuit.

(c) The emergency stop function shall override all other functions and operations in all modes.

(d) The emergency stop circuit shall:

(1) operate by deactivation or deenergization and on loss of the electrical supply.

(2) have only hardware based components (i.e. it shall not rely on software to operate) although it may be possible to initiate the circuit from the software of the programmable electronic system.

(3) signal the programmable electronic system that an emergency stop has been executed.

(e) Where the affected machine is associated with other machines working in a coordinated manner that have individual programmable electronic systems or a supervisory programmable electronic system, provisions shall be made to signal the other programmable electronic systems that an emergency stop has been executed.

(f) Each machine shall have provisions to connect external emergency stop devices, safeguards, or interlocks to the emergency stop circuit.

(g) Where a separate emergency stop device is provided, it shall be necessary to reset the emergency stop circuit manually before any machine motion may be initiated. The resetting of the emergency stop circuit by itself shall not initiate any motion.

7-7 Zone Stop.
The zone stop circuit shall function as a controlled stop followed by removal of power to the machine actuators that can cause a hazardous condition.

The zone stop circuit shall override all other functions and operations in all modes except emergency stop. The zone stop circuit may be either hardware or software based. It can be initiated from a hardware device, the logic or software of the machine controller, or over a communications network or link. It shall signal the logic or software that such a condition exists. For a group of machines working in a coordinated manner and having more than one machine controller, provisions shall be made to signal the other controllers that a zone stop has been executed.

Provisions to connect safeguards, zone sensors (e.g. presence sensing devices), or interlocks to this circuit shall be provided. It shall be necessary to reset the power to the machine actuators before any operation that could result in a hazardous condition is initiated. The resetting of the power to the machine actuators by itself shall not initiate any operation.

7-8 Normal Stop.

(a) The normal stop(s) shall function as a controlled stop.

(b) The stop circuit(s):

(1) shall override all other functions and operations in all modes except emergency stop.

(2) shall be permitted to be either hardware or software based.

(3) shall be permitted to be initiated from a hardware device, the logic or software of the machine programmable electronic system, or over a communications network or link

(4) shall signal the logic or software that a normal stop has been executed.

(c) Where the affected machine is associated with other machines working in a coordinated manner that have individual programmable electronic systems or a supervisory programmable electronic system, provisions shall be made to signal the other programmable electronic systems that a normal stop has been executed.

7-9 Start Circuits.

(a) The start of a cycle or operation shall only be possible where all the safety measures for personnel, the machine, and the work in progress are fulfilled.

(b) Suitable interlocks shall be provided to secure correct sequential starting of cycles and operations.

7-10 Hold to Run Circuits. Where used, hold to run circuits [e.g., jog, inch circuit(s)], shall be designed to require continuous actuation of the control device(s) to achieve operation (i.e., machine motion) particularly when a hazardous condition is present.

7-11 Operating Modes. Each machine shall be permitted to have one or more operating modes (e.g., teach for robots) that are determined by the type of machine and its application.

Where hazardous conditions can arise from mode selection, such operation shall be protected by suitable means (e.g., key operated switch, access code). Mode selection by itself shall not initiate operation. A separate action by the operator shall be required.

Safeguards shall remain effective for all operating modes.

7-12 Feed Interlocked with Spindle Drive. Interlocking shall be provided so that the spindle drive

motor controller is activated before the tool is driven into the wrkpiece.

7-13 Machinery Door Interlocking. Hinged or sliding doors providing ready access to compartments containing belts, gears, or other moving parts that may expose hazardous conditions shall be interlocked through limit switches or other means to prevent operation of the equipment when the doors are not closed.

7-14 Motor Contactors and Starters. Motor contactors and starters that initiate opposing motion shall be both mechanically and electrically interlocked to prevent simultaneous operation.

7-15 Relays and Solenoids. Relays and solenoids which are mechanically interlocked shall be electrically interlocked.

7-16 Set-Up Mode. Where necessary for setup purposes and when under supervised control, interlocks shall be permitted to be bypassed by qualified personnel provided that other protective interlocks for the safety of personnel shall remain operational.

7-17 Two-Hand Control Circuits. Where used to initiate potentially hazardous motion, two-hand control circuits shall:

(a) be protected against unintentional operation.

(b) have the pushbutton contacts connected in series and shall be arranged by design and construction or separation, or both, to require the concurrent use of both hands to initiate the machine operation.

(c) incorporate an antirepeat feature for machines that would present a hazard if an unintended repeat cycle occurred.

NOTE: See ANSI B11 series standards

Chapter 8   Control Equipment

8-1   Connections.   Means for making conductor connections shall be provided on or adjacent to all control devices mounted in the control enclosure.

8-2   Subpanels.   Subpanels with concealed or inaccessible internal wiring or devices shall be mounted and wired so as to be removable.

8-3   Manual   and   Electro-Mechanical   Motor Controllers.

(a) Each motor controller shall be identified and shall be capable of starting and stopping the motor(s) it controls and, for alternating current motors, shall be capable of interrupting the stalled rotor current of the motor(s). Controllers rated in horsepower shall be used for motors rated 1/8 horsepower or larger. The motor controller shall be sized in accordance with Table 8-3(a).

Exception: Other motor controllers shall be permitted provided that they are identified as suitable for the intended use and protected in accordance with the marked ratings.

NOTE: See definition of Identified in Appendix A.

Table 8-3(a) Horsepower Ratings for Three-Phase, Single-Speed Full Voltage Magnetic Controllers for Nonplugging and Nonjogging Duty

Size of Motor Controller	Service-Limit Current Rating Amperes*	Three-Phase Horsepower at		
		200 Volts	230 Volts	460/575 Volts
00	11	1 1/2	1 1/2	2
0	21	3	3	5
1	32	7 1/2	7 1/2	10
2	52	10	15	25
3	104	25	30	50
4	156	40	50	100
5	311	75	100	200
6	621	150	200	400
7	932	-	300	600
8	1400	-	450	900
9	2590	-	800	1600

Reference ANSI/NEMA ICS-2-1988, Table 2-321-1.
*The service-limit current ratings shown in Tables 8-3(a) and 8-3(c) represent the maximum rms current, in

amperes, the controller may be expected to carry for protracted periods in normal service.

(b) Alternating current motor controllers shall open all of the supply conductors leading to associated motors.

(c) Where machine operation requires a motor controller to repeatedly open high motor current, such as in plug-stop, plug-reverse, or jogging (inching) duty, requiring continuous operation with more than five openings per minute, the controller shall be de-rated in accordance with Table 8-3(c).

Exception: Other motor controllers shall be permitted provided they are identified as suitable for the intended use and protected in accordance with marked ratings,

NOTE: See definition of Identified in Appendix A.

Table 8-3(c) Horsepower Ratings for Three-Phase, Single-Speed, Full Voltage Magnetic Controllers for Special Duty Applications

Size of Controller	Continuous Current Rating* Amperes	Horsepower at 60 Hertz			Service-Limit Current Rating** Amperes
		200 Volts	250 Volts	460 or 575 Volts	
0	18	1 ½	1 ½	2	21
1	27	3	3	5	32
2	45	7 ½	10	15	52
3	90	15	20	30	104
4	135	25	30	60	156
5	270	60	75	150	311
6	540	125	150	300	621
9	2250		800	1600	2590

Reference ANSI/NEMA ICS 2-1988, Table 2-321-3

NOTE: Refer to ANSI/NEMA ICS-2 1988 for horsepower ratings of single-phase, reduced voltage, or multispeed motor controller application.

*The continuous-current ratings shown in Table 8-3(a) and 8-3(c) represent the maximum rms current, in amperes, the controller may be expected to carry continuously without exceeding the temperature rises permitted by Part ICS 1-109 of NEMA Standards Publication No. ICS 1.
**The service-limit current ratings shown in Tables 8-3(a) an 8-3(c) represent the maximum rms current, in ampere, the controller may be expected to carry for protracted periods in normal service. At service-limit

current ratings, temperature rises may exceed those obtained by testing the controller at its continuous current rating. The current rating of overload relays or the trip current of other motor protective devices used shall not exceed the service-limit current rating of the controller.

(d) Several motors shall be permitted to be operated from one motor controller where separate overload protection is provided for each motor, and the horsepower rating of the controller is not exceeded.

8-4 Marking on Motor Controllers. A controller for a motor rated 1/8 horsepower or more shall be marked with the voltage, phase, horsepower rating, and such other data as may be needed to properly indicate the motor for which it is suitable.

Chapter 9   Control Enclosures and Compartments

9-1 Type.

(a) Enclosures and compartments shall be nonventilated with construction and sealing suitable for the intended environment.

Exception: Equipment requiring ventilation shall be permitted to be:
> (1) housed in a separate ventilated portion of the enclosure or compartment, or
> (2) housed in a separate ventilated enclosure or compartment.

(b) Ventilated enclosures and compartments shall be constructed to prevent the entrance of any deleterious substance normal to the operating environment and shall prevent the escape of sparks or burning material.

9-2 Nonmetallic Enclosures.
Nonmetallic enclosures identified for the purpose shall be permitted. For grounding provisions, see Section 17-3.

9-3 Compartment Location.
Compartments for built-in control shall be completely isolated from coolant and oil reservoirs. The compartment shall be readily accessible and completely enclosed; it shall not be considered enclosed where it is open to the floor, the foundation upon which the machine rests, or other compartments of the machine that are not clean and dry.

9-4 Wall Thickness.
The walls of compartments shall be not less than the following .0625 in. (1.5 mm) for sheet steel; 1/8 in. (3.2 mm) for cast metal, or 3/32 in. (2.38 mm) for malleable iron.

9-5 Dimensions.
The depth of the enclosure or compartment including doors or covers shall not be less than the maximum depth of the enclosed equipment plus the required electrical clearances.

9-6 Doors.
Enclosures or compartments shall have one or more hinged doors that shall swing about a vertical axis and shall be held closed with captive fasteners or vault-type hardware. The thickness of metallic doors shall not be less than that indicated in Section 9-4. The width of doors shall not exceed 40 in. (1016 mm).

Exception: Where the motor(s) on the machine totals two horsepower or less, covers held on with captive screw-type fasteners shall be permitted.

9-7 Gaskets.
Where gaskets are used they shall be of an oil-resistant material and shall be securely attached to the door or enclosure.

9-8 Interlocks.
Any door(s) that permits access to live parts operating at 50 volts or more shall be so interlocked that the door(s) cannot be opened unless all power is disconnected.

Exception No. 1: External interlocking circuits operating at less than 150 volts need not be disconnected provided that the circuit conductors are identified with a yellow-colored insulation as described in Section 14-1(a) and a warning marking is attached to the door in accordance with section 2-5(a).

Exception No. 2: It shall be permitted to provide a means for qualified persons to gain access without removing power. The interlocking shall be reactivated automatically when the door(s) is closed.

Exception No. 3: Where an attachment plug is used as the disconnecting means and a warning marking is attached to the door in accordance with Section 2-5(b).

Exception No. 4: Where the motor(s) on the machine totals two horsepower or less, an external, non-interlocked disconnecting means shall be permitted provided that the disconnecting means is within sight and readily accessible, the control enclosure door or cover is marked with a warning indicating that the power shall be removed by the disconnecting means before the enclosure is opened, and further provided that a tool is required to open the enclosure.

9-9 Interior Finish.
The interior of control enclosures and exposed surfaces of panels mounted therein shall be finished in a light color.

Exception: An enclosed motor controller for a single motor.

9-10 Warning Mark.
All control enclosures and compartments which do not clearly show that they contain electrical devices and that are not electrically interlocked shall be marked with a black or red lightning flash on a yellow background within a black or red triangle as shown in Appendix C, Section C-1.

9-11 Print Pocket. A print pocket sized to accommodate the electrical diagrams shall be attached to the inside of the door of the control enclosure or compartment. Single door and multidoor enclosures shall have at least one print pocket.

Chapter 10   Location and Mounting of Control Equipment

10-1 General Requirements.

(a) Control equipment shall be mounted and located so that it will not interfere with machine adjustments or maintenance.

(b) Pipelines, tubing, or devices for handling air, gases, or liquids shall not be located in enclosures or compartments containing electrical control equipment.

Exception: Equipment for cooling electronic devices.

10-2 Control Panels.

(a) All devices connected to the supply voltage, shall be grouped devices connected only to control voltages.

Exception: Where the supply voltage is 150 volts or less.

(b) Terminal blocks for power circuits shall be grouped separately from control circuits.

Exception: Grouped power terminals shall be permitted to be mounted adjacent to grouped control terminals.

(c) Terminal blocks shall be mounted to provide unobstructed access to the terminals and their conductors.

(d) The stationary part of each multiconductor control cable plug/receptacle shall be mounted to provide unobstructed access.

(e) The panel shall not be set to such depth from the door frame or other projecting portion of machine as to interfere with inspection or servicing.

(f) Starters, contactors, and other control devices shall be front-mounted on a rigid metal panel. Equipment shall be mounted so that any component can be replaced without removing the panel. No components shall be mounted behind door pillars unless adequate space is provided for replacement and servicing.

(g) Test points, where provided, shall be mounted to provide unobstructed access, plainly marked to correspond to the markings on the drawings, adequately insulated, and sufficiently spaced for connection of test leads.

10-3 Subpanels and Electronic Subassemblies.
Subpanels and electronic subassemblies mounted on the control panel or on other supporting means, e.g., rack or shelf, shall be mounted so that adequate space is provided for replacement and servicing.

10-4 Control Enclosure.
The enclosure shall be mounted in a manner and position to guard it against oil, dirt, coolant, and dust, and to minimize the possibility of damage from floor trucks or other moving equipment.

10-5 Clearance in Enclosures and Compartments.

(a) Enclosures or compartments for mounting control panels shall provide adequate space between panel and case for wiring and maintenance.

(b) Exposed, nonarcing, bare, live parts within an enclosure or compartment, including conduit fittings, shall have an air space between them and the uninsulated walls of the enclosure or compartment, of not less than 1/2 in. (12.7 mm). Where barriers between metal enclosures or compartments and arcing parts are required, they shall be of flame-retardant, noncarbonizing, insulating materials.

10-6 Machine Mounted Control Equipment.

a) Control equipment, such as limit switches, brakes, solenoids, position sensors, etc., shall be mounted rigidly in a reasonably dry and clean location, shall be protected from physical damage, and shall be free from the possibility of accidental operation by normal machine movements or by the operator. Such equipment shall be mounted with sufficient clearance from surrounding surfaces to make its removal and replacement easy and shall have a suitable enclosure for the termination of conduit as well as provisions for making electrical connections.

Exception No. 1: A solenoid sealed in an individual oil-filled container shall be permitted.

Exception No. 2: Prewired devices, such as limit switches, proximity switches, etc., provided with an identified cable need not be equipped with provisions for termination of conduit.

(b) All limit switches or position sensors shall be installed so that accidental overtravel by the machine will not damage the limit switch or sensor.

(c) Solenoids for operating devices shall be mounted so that liquids shall drain away from the electrical component enclosure.

10-7 Rotary Control Devices. Devices such as potentiometers and selector switches having a rotating member shall be mounted to prevent rotation of the stationary member. Friction alone is not sufficient.

Chapter 11   Operator's Control Stations and Equipment

11-1   Pushbuttons, Selector Switches, Indicating Lights.

(a) All pushbutton and selector switch operators, indicating (pilot) lights, and illuminated pushbuttons shall be of the oiltight type.

Exception: Machines identified for the environment.

(b) Pushbutton operators, indicating (pilot) light lenses, and illuminated pushbutton lenses shall be color coded in accordance with Table 11-1.

    (1) The color RED shall be used for Stop, Emergency Stop, or Off operators only.

    (2) The preferred color of Start or On operators is GREEN except that BLACK, WHITE, or GRAY shall be permitted.

    (3) Pushbuttons that, when pressed, act alternately as Start and Stop or On and Off shall be BLACK, WHITE, or GRAY. RED or GREEN shall not be used.

    (4) Pushbuttons that cause movement when pressed and stop movement when they are released (e.g., jogging) shall be BLACK, WHITE, GRAY, or BLUE with a preference for BLACK.

    (5) Reset pushbuttons shall be BLUE, BLACK, WHITE, or GRAY except when they also act as a Stop or Off button in which case they shall be RED.

Exception: Stop function operators of the wobble-stick or rod-operated types in the bottom of a pendant station need not be colored red.

(c) Emergency Stop pushbutton operators shall be of the palm or mushroom type.

(d) Pushbutton operators used to initiate a stop function shall be of the extended operator or mushroom head types.

(e) Pushbutton operators used to initiate a start function or movement of machine elements (slides, spindles, carriers, etc.) shall be constructed or mounted so as to minimize inadvertent operation.

Exception: Mushroom-type operators shall be permitted to initiate start functions when installed in accordance with Section 17-7.

11-2 Emergency Stop Controls.

(a) Emergency Stop pushbuttons shall be located at each operator control station and at other operating stations where emergency shutdown shall be required.

(b) Stop and Emergency stop pushbuttons shall be continuously operable from all control and operating stations where located.

11-3 Foot-Operated Switches.

(a) Foot-operated switches shall be protected so as to prevent accidental actuation by falling or moving objects and from unintended operation by accidental stepping onto the switch.

11-4 Control Station Enclosures. All operator control station enclosures shall be dusttight, moisturetight, and oiltight.

Exception: Non-oiltight control station enclosures shall be permitted on machines where suitable for the environment.

11-5 Arrangement of Control Station Components. All Start pushbuttons shall be mounted above or to the left of their associated Stop pushbuttons.

Exception No. 1: Start pushbuttons in series, such as operating pushbuttons on punch presses.

Exception No. 2: Wobble-stick or rod-operated emergency stop pushbuttons mounted in the bottom of pendant stations.

11-6 Legends. A legend shall be provided for each control station component to identify its function and shall be located so that it can be read easily by the equipment operator from the normal operator position. The legends shall be durable and suitable for the operating environment.

Table 11-1 Color Coding for Pushbuttons, Indicating (Pilot) Lights, and Illuminated Pushbuttons.

Color	Device Type	Typical Function	Examples
RED	Pushbutton	Emergency Stop, Stop, Off	Emergency Stop button, Master Stop button, Stop of one or more motors
	Pilot Light	Danger or alarm, abnormal condition requiring immediate attention.	Indication that a protective device has stopped the machine, e.g., overload
	Illuminated Pushbutton		Machine stalled because of overload, etc. (use of RED illuminated pushbutton shall not be permitted for emergency stop.
YELLOW (AMBER)	Pushbutton	Return, Emergency Return, Intervention-suppress abnormal conditions	Return of machine elements to safe position, override other functions previously selected. Avoid unwanted changes.
	Pilot Light	Attention, caution/ marginal condition. Change or impending change of conditions.	Automatic cycle or motors running; some value (pressure, temperature) is approaching its permissible limit. Ground fault indication. Overload that is permitted for a limited time.
	Illuminated Pushbutton	Attention or caution/Start of an operation intended to avoid dangerous conditions.	Some value (pressure, temperature) is approaching its permissible limit; pressing button to override other functions previously selected.
GREEN	Pushbutton	Start-On	General or machine start; start of cycle or partial sequence; start of one or more motors; start of auxiliary sequence; energize control circuits.
	Pilot Light	Machine Ready; Safety	Indication of safe condition or authorization to proceed. Machine ready for operation with all conditions normal or cycle complete and machine ready, to be restarted.
	Illuminated Pushbutton	Machine or Unit ready for operation/Start or on	Start or On after authorization by light; start of one or more motors for auxiliary functions; start or energization of machine elements.
BLACK	Pushbutton	No specific function assigned.	Shall be permitted to be used for any function except for buttons with the sole function of Stop or Off; inching or jogging

Color	Device Type	Typical Function	Examples
WHITE or CLEAR	Pushbutton	Any function not covered by the above.	Control of auxiliary functions not directly related to the working cycles.
	Pilot Light	Normal Condition Confirmation	Normal pressure, temperature.
	Illuminated Pushbutton	Confirmation that a circuit has been energized or function or movement of the machine has been started/Start-On, or any preselection of a function.	Energizing of auxiliary function or circuit not related to the working cycle; start or preselection of direction of feed motion or speeds.
BLUE or GRAY	Pushbutton, Pilot Light, or Illuminated Push-button	Any function not covered by the above colors.	

For illuminated pushbuttons, the function(s) of the light is separated from the function(s) of the button by a virgule (/).

11-7 Location of Control Stations.

(a) All stations shall be mounted in locations that will minimize exposure to oil, coolant, and other contaminants.

(b) Controls shall be within normal reach of the machine operator, and shall be so placed that the operator does not have to reach past spindles or other moving parts.

(c) Controls shall be located so that unintentional operation by normal movement of the machine, operator, or work will be unlikely.

11-8 Pendant Stations.

(a) Pendant operator control station enclosures shall be oiltight.

(b) A wobble stick or rod operator at the bottom of the station shall be permitted for Emergency Stop controls.

(c) Pendant pushbutton stations shall be supported by suitable means other than the flexible electrical conduit or multiconductor cable.

(d) Grounding and bonding shall comply with Sections 17-2, 17-3, 17-4, and 17-6.

Chapter 12   Accessories and Lighting.

12-1 Attachment Plugs and Receptacles External to the Control Enclosure.

(a) Attachment plug and receptacles shall be listed for the applied voltage and shall be of the locking type where rated greater 20 amps. Where used on 300 volts or more, they shall be skirted and constructed to contain any arc generated when a connection is made or broken.

(b) Attachment plugs and receptacles shall be provided with a grounding pole and so constructed that the grounding pole is made before any current-carrying poles are made, and is not broken until all current-carrying poles of the attachment plug have been disconnected. A grounding pole shall not be used as a current-carrying part.

(c) Attachment plugs and receptacles shall be designed to prevent the entrance of oil or moisture when in an operating position. Means shall be provided to cover the receptacle when the plug is removed.

Exception: Where temperatures require the use of high-temperature attachment plugs and receptacles.

12-2 Receptacles Internal to the Control Enclosure.

(a) Receptacles internal to the control enclosure shall be permitted only for the following:

1. Maintenance equipment.

2. AC power distribution within the enclosure to electronic assemblies designed and approved for cord and plug connection.

(b) Receptacles shall be of the parallel-blade grounding type rated at 125 volts, 15 amperes.

(c) Receptacles shall be supplied from a 120 volt ac source and shall have individual overcurrent protection not to exceed 15 amperes.

(d) Receptacles for maintenance equipment shall be separate from receptacles for other purposes and shall have individual overcurrent protection not exceeding 15 amperes.

(e) The source of power shall be the equipment control transformer, a separate isolating transformer, or, in the case of receptacles for maintenance equipment only, the maintenance lighting circuits as permitted in Sections 12-3(b)(3) and 12-3(b)(5).

(f) The receptacles shall not be accessible when the equipment doors or covers are in the closed position.

12-3 Control Panel, Instrument, and Machine Work Lights.

(a) The lighting circuit voltage shall not exceed 150 volts between conductors.

(b) Lights shall be supplied from one of the following sources:

1. A separate isolating transformer connected to the load side of the machine disconnecting means. Overcurrent protection shall be provided in the secondary circuit.

2. A grounded 120-volt control circuit with separate overcurrent protection for the lighting circuit.

3. The plant lighting circuit shall be permitted for the supply of a maintenance lighting circuit in control enclosures only.

4. Where the motor(s) on the machine totals two horsepower or less, it shall be permitted to connect the machine worklight to the plant lighting circuit.

5. A separate isolating transformer connected to the line side of the main disconnecting means shall be permitted for the supply of a maintenance lighting circuit in control enclosures only.

6. The line side of the main disconnecting means where a separate primary disconnecting means, isolating transformer, and secondary overcurrent protection are furnished in an enclosure and mounted within the control enclosure, adjacent to the main disconnecting means.

(c) The conductors to stationary lights used as an integral part of the machine shall be Type MTW, and the conductors within the fixtures shall not be smaller than No. 18 AWG.

(d) Flexible cords shall be Type SO, STO, STOW, or SJO, SJOW SJTO and shall not incorporate in-line switches.

(e) Grounding shall comply with the provisions of Section 17-8.

(f) Machine work lights shall not contain switches or receptacles where exposed to liquids or condensing mists unless identified for the purpose.

(g) Stroboscopic effects from light shall be avoided.

(h) Reflectors and protectors shall be supported by a bracket and not the lampholder.

Chapter 13　Conductors

Table 13-2(a) Single Conductor Characteristics

13-1 General.

(a) These requirements cover thermoplastic Type MTW, THHN, THW, and THWN 600-V wires and cables for use as specified in the NFPA 70. *National Electrical Code*, UL 1063-1990 *Machine-Tool Wires and Cables* and ASTM Standards B 8-86, *Concentric-Lay-Stranded Copper Conductors, Hard, Medium-Hard, or Soft.* B 33-81, (R-1985), *Tinned Soft or Annealed Copper Wire for Electrical Purposes,* B 174-71 (R-1985), *Bunch-Stranded Copper Conductors for Electrical Conductors*, and B 286-89, *Copper Conductors for Use in Hookup Wire for Electronic Equipment..* The insulation and the finished wires and cables shall be suitably flame retardant and have temperature limits and characteristics as listed below:

1. MTW - Moisture-, Heat-, and Oil-Resistant Thermoplastic

60°C (140°F) Wet Locations

90°C (194°F) Dry Locations

2. THHN - Heat-Resistant Thermoplastic

90°C (194°F) Dry Locations

3. THW - Moisture- and Heat-Resistant Thermoplastic

75°C (167°F) Dry and Wet Locations

4. THWN-Moisture- and Heat-Resistant Thermoplastic

75°C (167°F) Dry and Wet locations

(b) Multiconductor flexible cords, Type SO, STO, STOW, or SJO, SJOW SJTO shall be permitted.

(c) Mineral-insulated (metal-sheathed) cable, Type MI, shall be permitted. Temperature range-85°C (185°F) Dry and Wet Locations.

13-2 Conductors.

(a) Conductors of AWG sizes 22 through 4/0 and MCM sizes 250 through 1000 shall be only of stranded soft-annealed copper. Conductor cross-sectional area, dc resistance, and stranding is listed in Table 13-2(a).

Size (AWG/ Kcmil)	Cross-Sectional Area Nominal (CM/mm²)	DC Resistance at 25°C (ohms/1000 ft)	Minimum Stranding Nonflexing	Number (ASTM Class) Flexing	Constant Flex (ASTM Class/AWG Size)
22AWG	640/0.324	17.2	7(')	7(')	19(M/34)
20	1020/0.519	10.7	10(K)	10(K)	26(M/34)
18	1620/0.823	6.77	16(K)	16(K)	41(M/34)
16	2580/1.31	4.26	19(C)	26(K)	65(M/34)
14	4110/2.08	2.68	19(C)	41(K)	41(K/30)
12	6530/3.31	1.68	19(C)	65(K)	65(K/30)
10	10380/5.261	1.060	19(C)	104(K)	104(K/30)
8	16510/8.367	0.6663	19(C)	(\)	(-)
6	26240/13.30	0.4192	19(C)	(\)	(-)
4	41740/21.15	0.2636	19(C)	(\)	(-)
3	52620/26.67	0.2091	19(C)	(\)	(-)
2	66360/33.62	0.1659	19(C)	(\)	(-)
1	83690/42.41	0.1315	19(B)	(\)	(-)
1/0	105600/53.49	0.1042	19(B)	(\)	(-)
2/0	133100/67.43	0.08267	19(B)	(\)	(-)
3/0	167800/85.01	0.06658	19(B)	(\)	(-)
4/0	211600/107.2	0.05200	19(B)	(\)	(-)
250kcmil	/127	0.04401	37(B)	(\)	(-)
300	/152	0.03667	37(B)	(\)	(-)
350	/177	0.03144	37(B)	(\)	(-)
400	/203	0.02751	37(B)	(\)	(-)
450	/228	0.02445	37(B)	(\)	(-)
500	/253	0.02200	37(B)	(\)	(-)
550	/279	0.02000	61(B)	(\)	(-)
600	/304	0.01834	61(B)	(\)	(-)
650	/329	0.01692	61(B)	(\)	(-)
700	/355	0.01572	61(B)	(\)	(-)
750	/380	0.01467	61(B)	(\)	(-)
800	/405	0.01375	61(B)	(\)	(-)
900	/456	0.01222	61(B)	(\)	(-)
1000	/507	0.01101	61(B)	(\)	(-)

(A,B,) ASTM Class designation B and C per ASTM B 8-86.

(K) Class Designation K per ASTM B 174-71 (R-1980).

(') A class designation has not been assigned to this conductor but is designated as size 22-7 in ASTM B 286-74 (R-1979) and is composed of strands 10 mils in diameter (No. 30 AWG).

(\) Nonflexing construction shall be permitted for flexing service.

(-) Constant flexing cables are not constructed in these sizes.

Exception No. 1: Conductors with insulation characteristics consistent with the provisions of this chapter but with stranding other than that specified in Table 13-2(a) shall be permitted on individual devices that are purchased completely wired (i.e., motor starters, etc.).

Exception No. 2: Conductors subject to temperatures, voltages, environmental conditions, or flexing exceeding the ratings listed in this chapter shall have suitable characteristics.

(b) Where constant flexing service is required, conductor stranding shall conform to 13-2(a).

(c) Solid conductors AWG 24-30 of soft-annealed copper shall be permitted for use within control enclosures when not subject to flexing.

(d) Printed wire assemblies of flame-retardant material shall be permitted in place of conductor assemblies provided they are within control enclosures and are mounted in such a way as to minimize flexing or stress.

(e) Shielded conductors shall consist of stranded, annealed copper of 25 AWG or larger for single conductors used in subassemblies and 22 AWG or larger for all other uses. The conductors shall be in accordance with Tables 13-2(a) and 13-4(c) and shall have a metallic shield and an oil- and moisture-resistant outer covering such as vinyl plastic.

(f) Special conductors such as RG-/U transmission cable shall be permitted where necessary for the proper functioning of the equipment.

13-3 Conductor Sizing.

Conductors shall not be smaller than:

(a) Power circuits ....No. 14

(b) Lighting and control circuits on the machine and in raceways... ....No. 16

Exception: In a jacketed, multiconductor cable assembly, No. 18 shall be permitted.

(c) Control circuits within enclosures or operator stations...No. 18

(d) Electronic, programmable Control 1/0; and static control:

(1) Conductors in raceways...............................No. 24

Exception: In a jacketed, multiconductor cable assembly No. 30 or larger shall be permitted.

(2) Conductors within control enclosures...........No. 26

Exception: For jumpers and special wiring applications (e.g. solderless wrap or wire-clip type connections or shielded conductors), conductors No. 30 or larger shall be permitted.

13-4 Wire Insulation.

(a) Where "thermoplastic" or "PVC" appears in this standard, the intention is to designate a synthesized compound whose characteristic constituent is polyvinyl chloride or a copolymer of vinyl chloride or vinyl acetate.

Every wire requiring insulation and intended for use as a single conductor or in a cable shall be insulated for its entire length with properly compounded homogeneous PVC material.

(b) The color of the insulation shall be solid, or it shall have one or more stripes of different colors.

(c) The average and the minimum thickness of the insulation in constructions A and B shall be in accordance with Table 13-4(c).

(d) Construction B shall have a nylon jacket applied directly over the insulation. The jacket shall be snug on the insulation and shall be at least as thick as indicated in Table 13-4(c).

(e) Wire insulation shall be identified and adequate for the voltage on that conductor. Where the conductors are run with or adjacent to other conductors, all conductors shall have insulation rated for the maximum voltage involved.

Exception: Bare conductors such as capacitor or resistor leads, jumpers between terminals, etc., shall be permitted where the method of securing provides adequate electrical clearance.

Table 13-4(c) Single Conductor Insulation Thickness of Insulation in Mils*
[Average/Minimum(Jacket)]

Wire Size	A	B
22AWG	30/27	15/13(4)
20	30/27	15/13(4)
18	30/27	15/13(4)
16	30/27	15/13(4)
14	30/27	15/13(4)
12	30/27	15/13(4)
10	30/27	20/18(4)
8	45/40	30/27(5)
6	60/54	30/27(5)
4-2	60/54	40/36(6)
1-4/0	80/72	50/45(7)
250-500MCM	95/86	60/54(8)
550-1000	110/99	70/63(9)

(*) UL 1063 Table 1.1 *NEC* Construction
A-No outer covering
B-Nylon Covering

13-5 Conductor Ampacity.

(a) The continuous current carried by conductors shall not exceed the values given in Table 13-5(a).

Table 13-5(a) Conductor Ampacity Based on Copper Conductors with 60°C and 75°C Insulation in an Ambient Temperature of 30°C

Conductor Size AWG	Ampacity In Cable or Raceway		Control Enclosure
	60°C	75°C	60°†
30	-	0.5	0.5
28	-	0.8	0.8
26	-	1	1
24	2	2	2
22	3	3	3
20	5	5	5
18	7	7	7
16	10	10	10
14	15	15	20
12	20	20	25
10	30	30	40
8	40	50	60
6	55	65	80
4	70	85	105
3	85	100	120
2	95	115	140
1	110	130	165
0	125	150	195
2/0	145	175	225
3/0	165	200	260
4/0	195	230	300
250	215	255	340
300	240	285	375
350	260	310	420
400	280	335	455
500	320	380	515
600	355	420	575
700	385	460	630
750	400	475	655
800	410	490	680
900	435	520	730
1000	455	545	780

† Sizing of conductors in wiring harnesses or wiring channels shall be based on ampacity for cables.
NOTE 1: Wire types listed in Section 13-1 shall be permitted to be used at the ampacities listed in this table.
NOTE 2: For ambient temperatures other than 30°C, see *NEC* table 310-16 correction factors.

NOTE 3: The sources for the ampacities in this table are Tables 310-16 and 310-17 of the *NEC*.

(b) Motor circuit conductors shall have an ampacity not less than 125 percent of the full-load current rating of the highest rated motor in the group plus the sum of the full load current ratings of all other connected motors and apparatus in the group which may be in operation at the same time.

(c) Combined load conductors shall have an ampacity not less than 125 percent of the full-load current rating of all resistance heating loads plus 125 percent of the full-load current rating of the highest rated motor plus the sum of the full-load current ratings of all other connected motors and apparatus which may be in operation at the same time.

(d) The maximum size of a conductor as selected from Table 13-5(a) and connected to a motor controller shall not exceed the values given in Table 13-5(d).

Exception: Where other motor controllers are used, the maximum conductor size shall not exceed that specified by the manufacturer.

Table 13-5(d) Maximum Conductor Size for Given Motor Controller Size*

Motor Controller Size	Maximum Conductor Size AWG or MCM
00	14
0	10
1	8
2	4
3	0
4	000
5	500

*See ANSI/NEMA ICS 2-1983 Table 2, 110-1

13-6 Wire Markings.

(a) A durable legend printed on the outer surface of the insulation of construction A, on the outer surface of the nylon jacket of construction B, on the outer surface of the insulation under the jacket of construction B (only if clearly legible through the nylon), or on the outer surface of the jacket of a multiconductor cable shall be repeated at intervals of no more than 24 inches (610 mm) throughout the length of the single-conductor or the multiconductor cable.

Exception: Size smaller than Number 18 shall be permitted to be marked on the reel or smallest unit of shipping carton.

(b) The legend shall include the manufacturer's name or trademark, the wire type, voltage rating, and gage or size.

(c) Where the conductor size is AWG 16-10 and the stranding is intended for flexing service, the legend shall include "flexing" or "Class K".

Chapter 14   Wiring Methods and Practices

14-1 General Requirements.

(a) Conductors shall be identified at each termination to correspond with the identification on the diagrams and shall be color coded as follows:

BLACK - Line, load, and control circuits at line voltage.

RED - AC control circuits at less than line voltage.

BLUE - DC control circuits.

YELLOW - Interlock control circuits supplied from an external power source.

NOTE: The international and European standards require the use of the color ORANGE for this purpose. *(See IEC 204-1 for specific requirements).*

GREEN (with or without one or more yellow stripes) - Equipment grounding conductor where insulated or covered.

NOTE: The international and European standards require the use of the bi-color GREEN-AND-YELLOW for this purpose.*(See IEC 204-1 for specific requirements.)*

WHITE or NATURAL GRAY - Grounded circuit conductor.

NOTE: The international and European standards require the use of the color LIGHT BLUE for this purpose.*(See IEC 204-1 for specific requirements.)*

Exception No. 1: Internal wiring on individual devices purchased completely wired.

Exception No. 2: Where insulation is used that is not available in the colors required.

Exception No. 3: Where multiconductor cable is used.

Exception No. 4: Conductors used to connect electronic, precision, static, or similar devices or panels.

Exception No. 5: Where local conditions require that the control circuit be grounded, it shall be sufficient to use a green (with or without one or more yellow stripes) or a bare conductor from the transformer terminal to a grounding terminal on the control panel.

Exception No. 6: Additional colors shall be permitted to be used to facilitate identification between control panels and devices on the equipment; however, black shall be used for all wiring at line voltage.

(b) Conductors and cables shall be run without splices from terminal to terminal.

Exception: Splices shall be permitted to leads attached to electrical equipment, such as motors and solenoids, and shall be insulated with oil-resistant electrical tape or insulation equivalent to that of the conductors.

(c) Terminals on terminal blocks shall be plainly identified to correspond with markings on the diagrams.

(d) Shielded conductors shall be so terminated to prevent fraying of strands and to permit easy disconnection.

(e) Identification tags shall be made of oil-resistant material. Where wrap-type adhesive strips are used, they shall be of a length not less than twice the circumference of the wire. Sleeve-type tags shall be applied so they will not slip off the wire.

(f) Terminal blocks shall be wired and mounted so that the internal and external wiring does not cross over the terminals. Not more than two conductors shall be terminated at each terminal connection.

Exception: More than two conductors shall be permitted where the terminal is identified.

14-2 Panel Wiring.

(a) Panel conductors shall be supported where necessary to keep them in place. Wiring channels shall be permitted where made of a flame-retardant insulating material.

(b) Where back connected control panels are used, access doors or swingout panels that swing about a vertical axis shall be provided.

(c) Multiple-device control panels shall be equipped with terminal blocks or with attachment plugs and receptacles for all outgoing control conductors.

14-3 Machine Wiring.

(a) Conductors and their connections external to the control panel enclosure shall be totally enclosed in suitable enclosures or raceways as described in Chapter 15, unless otherwise permitted in this section.

(b) Fittings used with raceways or multiconductor cable shall be liquidtight.

Exception: Liquidtight fittings are not required where flexible metal conduit is permitted by Exception to Section 14-3(d).

(c) Liquidtight flexible conduit or multiconductor cable shall be used where necessary to employ flexible connections to pendant pushbutton stations. The weight of pendant stations shall be supported by chains or wire rope external to the flexible conduit or multiconductor cable.

(d) Liquidtight flexible conduit or multiconductor cable shall be used for connections involving small or infrequent movements. They shall also be permitted to complete the connection to normally stationary motors, limit switches, and other externally mounted devices.

Exception: Where subjected to temperature exceeding the limits for liquidtight flexible metal conduit, flexible metal conduit shall be permitted.

(e) Connections to frequently moving parts shall be made with conductors for flexing service as shown in Table 13-2(a). Flexible cable and conduit shall have vertical connections and shall be installed to avoid excessive flexing and straining.

Exception: Horizontal connections shall be permitted where the flexible cable or conduit is adequately supported.

(f) Where flexible conduit or cable is adjacent to moving parts, the construction and supporting means shall prevent damage to the flexible conduit or cable under all conditions of operation.

Exception: Prewired devices such as limit switches, proximity switches, etc., provided with an identified cable need not be provided with provisions for termination of conduit.

(g) All conductors of any ac circuit shall be contained in the same raceway.

(h) Conductors connected in ac circuits and conductors connected in dc circuits shall be permitted in the same raceway regardless of voltage, provided they are all insulated for the maximum voltage of any conductor in the raceway.

(i) Connection through a polarized grounding-type attachment plug and receptacle shall be permitted where equipment is removable. The male plug shall be connected to the load circuit.

(j) Where construction is such that wiring must be disconnected for shipment, terminal blocks in an accessible enclosure or attachment plugs and receptacles shall be provided at the sectional points.

(k) The installation of flexible conduit and cable shall be such that liquids will drain away from the fittings.

(l) Where liquidtight flexible metal conduit is used for flexible applications, fittings shall be identified.

14-4 Wire Connectors and Connections.

(a) Pressure connectors shall be used to connect conductors to devices with lug-type terminals that are not equipped with saddle straps or equivalent means of retaining conductor strands.

Exception No. 1: Solder connections shall be permitted to be used within the protective shell of a plug or receptacle and for internal connections of a subassembly that can be removed for bench service (See Section 14-4(b)).

Exception No. 2: Wire-wrapped connections shall be permitted to be used where circumstances permit and where applied by use of a tool specifically recommended for the purpose.

(b) Soldered connections shall conform to the following:

1. For manually soldered connections, rosin shall be used as flux.

2. Where printed circuit boards or other component assemblies are dip or wave soldered, special fluxes shall be permitted to be used following techniques developed specifically for these methods of fabrication.

3. All parts shall be pre-tinned before soldering unless the part is specifically plated to insure a good solder joint (e.g., MS-type connectors having gold plated contacts).

4. Each soldered connection shall be made with the least amount of solder that will assure a secure, high conductivity connection.

5. Insulation shall not be damaged by soldering.

6. Components which may be damaged by heat shall be suitably shielded from heat during soldering.

Chapter 15   Raceways, Junction Boxes and Pull Boxes

15-1 General Requirements.

(a) All sharp edges, flash, burrs, rough surfaces, or threads with which the insulation of the conductors may come in contact shall be removed from raceways and fittings. Where necessary, additional protection consisting of a flame-retardant, oil-resistant insulating material shall be provided to protect conductor insulation.

(b) Drain holes of 1/4 in. (6.4 mm) shall be permitted in raceways, junction boxes, and pull boxes subject to accumulations of oil or moisture.

NOTE: Raceways and junction boxes are provided for mechanical protection only. See Chapter 17 for acceptable means of equipment grounding.

15-2 Percent Fill of Raceways.
The combined cross-sectional area of all conductors and cables shall not exceed 50 percent of the interior cross-sectional area of the raceway. The fill provisions shall be based on the actual dimensions of the conductors and/or cables used.

15-3 Rigid Metal Conduit and Fittings.

(a) Rigid metal conduit and fittings shall be of galvanized steel meeting the requirements of ANSI Standards C80.1-1983 - *Specification for Rigid Steel Conduit, Zinc Coated*, and ANSI/NEMA FB 1-1983 (Rev. Sept. 1984), *Fittings and Supports for Conduit and Cable Assemblies*, or of a corrosion-resistant material suitable for the conditions. Dissimilar metals in contact which would cause galvanic action shall not be used. Conduit shall be protected against corrosion except at the threaded joints.

Exception: Threads at joints shall be permitted to be coated with an identified electrically conductive compound.

(b) Conduit smaller than 1/2 in. electrical trade size shall not be used.

(c) Conduit shall be securely held in place and supported at each end.

(d) Fittings shall be threaded unless structural difficulties prevent assembly. Where threadless fittings must be used, conduit shall be securely fastened to the equipment.

(e) Running threads shall not be used.

(f) Where conduit enters a box or enclosure, a bushing or fitting providing a smoothly rounded insulating surface shall be installed to protect the conductors from abrasion, unless the design of the box or enclosure is such to afford equivalent protection. Where conduit bushings are constructed wholly of insulating material, a locknut shall be provided both inside and outside the enclosure to which the conduit is attached.

Exception: Where threaded hubs or bosses that are an integral part of an enclosure provide a smoothly rounded or flared entry for conductors.

(g) Conduit bends shall be so made that the conduit can not be injured, and that the internal diameter of the conduit will not be effectively reduced. The radius of the curve of any field bend shall not be less than shown in Table 15-3(g).

Table 15-3(g) Minimum Radius of Conduit Bends

Size of Conduit (In.)	Radius of Bend Done by Hand (In.)[1]	Radius of Bend Done by Machine (In.)[2]
1/2	4	4
3/4	5	4 1/2
1	6	5 1/4
1 1/4	8	7 1/4
1 1/2	10	8 1/4
2	12	9 1/2
2 1/2	15	10 1/2
3	18	13
3 1/2	21	15
4	24	16
4 1/2	27	20
5	30	24
6	36	30

For SI units:(Radius) one inch = 25.4 millimeters

NOTE 1: For field bends done by hand, the radius is measured to the inner edge of the bend.

NOTE 2: For a single operation (one shot) bending machine designed for the purpose, the radius is measured to the center line of the conduit.

(h) A run of conduit shall contain not more than the equivalent of four quarter bends (360 degrees total).

15-4 Intermediate Metal Conduit. Intermediate metal (steel) conduit shall be permitted and shall be installed in conformance with the provisions of Section 15-3(b) through (h).

15-5 Electrical Metal Tubing. Electrical metallic (steel) tubing shall be permitted and shall be installed in conformance with the provisions of Sections 15-3(b) through (h).

15-6 Schedule 80 Rigid Nonmetallic Conduit.

(a) Rigid nonmetallic conduit Schedule 80 and fittings shall be listed.

(b) Conduit smaller that 1/2-in. electrical trade size shall not be used.

(c) Conduit shall be securely held in place and supported as follows:

Conduit Size (In.)	Maximum Spacing Between Supports (ft)
1/2 - 1	3
1 1/4 - 2	5
2 1/2 - 3	6
3 1/2 - 5	7
6	8

In addition, conduit shall be securely fastened within 3 ft. (914 mm) of each box, enclosure, or other conduit termination.

(d) Expansion fittings shall be installed to compensate for the thermal expansion and contraction. (Table 10, Chapter 9, *NEC*, 1990 edition.)

(e) Where a conduit enters a box or other fitting, a bushing or adaptor shall be provided to protect the wire from abrasion unless the design of the box or fitting is such as to provide equivalent protection.

(f) Conduit bends shall be so made that the conduit will not be injured and the internal diameter of the conduit will not be effectively reduced. The radius of the curve of any field bend shall not be less than shown in Table 15-3(g).

(g) A run of conduit shall not contain more than the equivalent of for quarter bends (360 degrees total).

(h) All joints between lengths of conduit, and between conduit and couplings, fittings, and boxes, shall be made with fittings designed for the purpose.

15-7 Liquidtight Flexible Metal Conduit and Fittings.

(a) Liquidtight flexible metal conduit shall consist of an oil-resistant, liquidtight jacket or lining in combination with flexible metal reinforcing tubing.

(b) Fittings shall be of metal and shall be designed for use with liquidtight flexible metal conduit.

(c) Liquidtight flexible metal conduit smaller than 3/8 in. electrical trade size shall not be used.

(d) Liquidtight flexible metal conduit shall be permitted to be of the extra flexible construction.

(e) Flexible conduit shall be installed in a manner that liquids will tend to run off the surface instead of draining toward the fittings.

15-8 Liquidtight Flexible Nonmetallic Conduit and Fittings.

(a) Liquidtight flexible nonmetallic conduit is a raceway of cross section of various types:

1. A smooth seamless inner core and cover bonded together and having one or more reinforcement layers between the core and cover.

2. A smoother inner surface with integral reinforcement within the conduit wall.

3. A corrugated internal and external surface with or without integral reinforcement within the conduit wall.

This conduit is oil-, and water-, and flame-resistant and, with fittings, is approved for the installation of electrical conductors.

(b) The conduit shall be resistant to kinking and shall have physical characteristics comparable to the jacket of multiconductor cable.

(c) The conduit shall be suitable for use at temperatures of 80°C and air and 60°C in the presence of water, oil, or coolant.

(d) Fittings shall be identified for their intended use.

(e) Liquidtight flexible nonmetallic conduit smaller than 3/8-in. trade size shall not be used.

(f) Flexible conduit shall be installed in such a manner that liquids will tend to run off the surface instead of draining toward the fittings.

15-9 Flexible Metal (Nonliquidtight) Conduit and Fittings.

(a) Flexible metal conduit shall consist of flexible metal tubing or woven wire armor.

(b) Fittings shall be of metal and shall be designed for use with flexible metal conduit.

(c) Flexible metal conduit smaller than 3/8-in. electrical trade size shall not be permitted.

Exception: Thermocouples and other sensors.

15-10 Wireways.

(a) Exterior wireways shall be permitted where rigidly supported and clear of all moving or contaminating portions of the machine.

(b) Metal thickness shall not be less than No. 14 MSG.

(c) Covers shall be shaped to overlap the sides; gaskets shall be permitted. Covers shall be attached to wireways by hinges or chains and held closed by means of captive screws or other suitable fasteners. On horizontal wireways, the cover shall not be on the bottom.

(d) Where the wireway is furnished in sections, the joints between sections shall fit tightly but need not be gasketed.

(e) Only openings required for wiring or for drainage shall be provided. Wireways shall not have unused knockouts.

15-11 Machine Compartments and Raceways.
Compartments or raceways within the column or base of a machine shall be permitted to enclose conductors, provided the compartment or raceway is isolated from coolant and oil reservoirs and is entirely enclosed. Conductors run in enclosed compartments and raceways shall be secured and so arranged that they will not be subject to physical damage.

15-12 Junction and Pull Boxes.
Junction and pull boxes shall not have unused knockout or openings and shall be constructed to exclude such materials as dust, flyings, oil, and coolant.

15-13 Motor Terminal Boxes.
Motor terminal boxes shall enclose only connections to the motor and motor-mounted devices, such as brakes, temperature sensors, plugging switches, tachometer generators.

Chapter 16   Motors and Motor Compartments

16-1 Access. Each motor and its associated couplings, belts, and chains shall be mounted where they are accessible for maintenance and not subject to damage.

16-2 Mounting Arrangement.

(a) The motor mounting arrangement shall be such that all motor hold-down bolts can be removed and replaced and terminal boxes reached. Unless bearings are permanently sealed, provision shall be made for lubricating the bearings. The motor nameplate shall indicate where permanently sealed bearings are used.

(b) Sufficient air circulation shall be provided so that the motor will not exceed its rated temperature rise at rated operating conditions.

(c) All motor-driven couplings, belts, and chains shall be easily replaceable.

(d) Pulley hubs on belted drives shall not extend beyond the end of the motor shaft.

16-3 Direction Arrow. Where reverse rotation can produce an unsafe condition, a direction arrow shall be installed. The arrow shall be adjacent to the motor and plainly visible.

16-4 Motor Compartments. Motor compartments shall be clean, dry, and adequately vented directly to the exterior of the machine. There shall be no opening between the motor compartment and any other compartment that does not meet the motor compartment requirements. Where a conduit or pipe is run into the motor compartment from another compartment not meeting the motor compartment requirements, any clearance around the conduit or pipe shall be sealed.

16-5 Marking on Motors. Motors shall be marked in accordance with Section 430-7 of NFPA 70, *National Electrical Code,* 1990 edition.

Chapter 17   Grounded Circuits and Equipment Grounding

17-1 General. This chapter applies to grounded circuits and the protective or grounding circuit of the equipment. The grounding circuit consists of conductors, structural parts of the electrical equipment, or both, that are all electrically connected or bonded together at a common point.

17-2 Grounding Conductors.

(a) Conductors used for grounding and bonding purposes shall be copper. Stipulations on stranding and flexing as outlined in this standard shall apply. [*See Sections 13-3 and 14-3(f)*].

(b) Grounding conductors shall be insulated, covered, or bare and shall be protected against physical damage. Insulated or covered grounding conductors shall be identified with a continuous green color with or without one or more yellow stripes.

Exception: It shall be permitted to use other color conductors provided the insulation or cover appropriately identified at all points of access.

(c) The minimum size of the grounding conductor shall be as shown in Table 17-2(c). Column "A" indicates maximum rating or setting of the overcurrent device in the circuit ahead of the equipment.

Table 17-2(c) Size of Grounding Conductors

Column "A" Amperes	Copper Conductor Size AWG
10	16* or 18*
15	14, 16* or 18*
20	12, 14*, 16* or 18*
30	10
40	10
60	10
100	8
200	6
300	4
400	3
500	2
600	1
800	0
1000	2/0
1200	3/0
1600	4/0

* Permitted only in multiconductor cable where connected to portable or pendant equipment.

(d) It is permissible to use machine members or structural parts of the electrical equipment in the grounding circuit provided that the cross-sectional area of these parts is at least electrically equivalent to the minimum cross-sectional area of the copper conductor required.

17-3 Equipment Grounding. The machine and all exposed, non-current carrying conductive parts, material, and equipment including metal mounting panels that are likely to become energized and are mounted in nonmetallic enclosures, shall be effectively grounded.

17-4 Exclusion of Switching Device. The grounding circuit shall not contain any switches or overcurrent protective devices. Links or plugs in the grounding circuit shall be permitted if properly labeled or interlocked with the control circuits.

17-5 Grounding Terminal. The entire grounding circuit or network shall be interconnected such that a single point for external connection will be conductively connected to all grounded parts. A terminal suitable for connecting an external grounding electrode conductor shall be provided at this point.

Exception: Where an attachment plug and receptacle are used as the disconnecting means, Section 5-11(f) shall apply.

17-6 Continuity of the Grounding Circuit.

(a) The continuity of the grounding circuit shall be ensured by effective connections through conductors or structural members.

(b) Bonding of equipment with bolts or other identified means shall be permitted where paint and dirt are removed from the joint surfaces or effectively penetrated.

(c) Moving machine parts, other than accessories or attachments, having metal-to-metal bearing surfaces shall be considered as bonded. Sliding parts separated by a non-conductive fluid under pressure shall not be considered as bonded.

(d) Portable, pendant, and resilient mounted equipment shall be bonded by separate conductors. Where multiconductor cable is used, the bonding conductor shall be included as one conductor of the cable.

(e) Raceways, wireways, and cable trays shall not be used as grounding or bonding conductors.

(f) When a part is removed, the continuity of the grounding circuit for the remaining parts shall remain intact.

17-7   Control Circuits.   Control circuits shall be permitted to be grounded or ungrounded. Where grounding is provided, that side of the circuit common to the coils shall be grounded at the control transformer if alternating current or at the power supply terminal if direct current.   For color coding of conductors see Section 14-1(a).

Exception No. 1 : Exposed control circuits as permitted by Section 7-2(a), Exception No. 2 shall be grounded.

Exception No. 2:   Overload relay contacts shall be permitted to be connected between the coil and the grounded conductor where the conductors between such contacts and coils of magnetic devices do not extend beyond the control enclosure.

17-8 Lighting Circuits.

(a) One conductor of all machine lighting and maintenance lighting circuits shall be grounded. The grounded conductor(s) shall be identified with a WHITE or NATURAL GRAY insulation.

(b) Where the lighting circuit is supplied by a separate isolation transformer, the grounding shall occur at the transformer. Where the equipment maintenance lighting circuit is supplied directly from the plant lighting circuit, the grounding shall occur at the grounding shall occur at the grounding terminal.

(c) The grounded conductor, where run to a screw-shell lampholder, shall be connected to the screw-shell.

Chapter 18   Electronic Equipment

18-1   General. This chapter applies to all types of electronic equipment including programmable electronic systems, subassemblies, printed circuit boards, electronic components, and other miscellaneous solid state equipment.

18-2 Basic Requirements.

(a) The provisions of Chapter 3 apply to electronic equipment.

(b) Subassemblies shall be readily removable or inspection or replacement.

(c) Transient suppression and isolation shall be provided where this equipment generates transient or electrical noise which can affect the operation of the equipment.

(d) Power supplies for electronic units that require memory retention shall have battery back-up of sufficient capacity to prevent memory loss for a period of at least 72 hours.

(e) Loss of memory contents shall prohibit the initiation of any hazardous conditions whose operation is dependant on memory contents.

(f) Outputs controlled by programmable electronic systems shall be protected from overload and short circuit conditions *(see Section 18-3(b) for grounding requirements).*

18-3 Programmable Electronic Systems.

(a) Programmable electronic systems shall be designed and constructed so that the ability to modify the application program shall be limited to authorized personnel and shall require special equipment or other means to access the program (e.g., access code, key operated switch).

Exception: For reasons of safety, the manufacturer or supplier shall be permitted to retain the right to not allow the user to alter the program.

(b) All input/output racks (remote or local), processor racks, and power supplies shall be electrically bonded together and connected to the grounding circuit *(see Section 17-2)* in accordance with the manufacturer's specifications. Where specified by the manufacturer, components and subassemblies shall be effectively bonded to the grounding circuit in accordance with the manufacturer's recommendations.

Chapter 19   Referenced Publications

19-1 The following documents or portions thereof are referenced within this document and shall be considered part of the requirements of this document. The edition indicated for each reference shall be the current edition as of the date of the NFPA issuance of this document.

19-2 NFPA Publication. National Fire Protection Association,1 Batterymarch Park, P. O. Box 9101, Quincy, MA 02269-9101.

NFPA 70-1990 *National Electrical Code*

19-3 ANSI Publications. American National Standards Institute, 1430 Broadway, New York, NY 10018

ANSI C80-1-1983, *Specifications for Rigid Steel Conduit, Zinc Coated*

ANSI C84.1-1982, *Voltage Ratings for Electrical Power Systems and Equipment (60Hz)*

ANSI Y32.2 1982, *Graphical Symbols for Electrical and Electronics Diagrams*

19-4   NEMA Publication.   National Electrical Manufacturers Association, 1201 L Street NW, Washington, DC 20037

NEMA FB 1-1983 (Rev. Sept. 1984), *Fittings and Supports for Conduit and Cable Assemblies*

Appendix A   Glossary of Terms

This Appendix is not a part of the requirements of this NFPA document, but is included for information purposes only.

Accessible (As Applied to Equipment). Admitting close approach: not guarded by locked doors, elevation, or other effective means. (NFPA 70, *National Electrical Code*, 1990 edition.)

Accessible, Readily (Readily Accessible). Capable of being reached quickly for operation, renewal, or inspections, without requiring those to whom ready access is requisite to climb over or remove obstacles or to resort to portable ladders, chairs, etc. (*See "Accessible."*) (NFPA 70, *National Electrical Code*, 1990 edition.)

Actuator (Machine). A power mechanism used to effect motion.

Adjustable Speed Drives. An electrical device or group of electrical devices that alters the drive motor output speed over a range in a controlled manner. This includes ac and dc voltage modes and frequency mode controls. Belt, chain, or roller shifting controllers are not included.

Alphanumeric. Pertaining to a character set that contains both letters and digits, but usually some other characters such as punctuation marks. (ANSI/IEEE Standard No. 100-88.)

Ampacity. The current in amperes a conductor can carry continuously under the conditions of use without exceeding its temperature rating. (NFPA 70, *National Electrical Code*, 1990 edition.)

Attachment Plug (Plug Cap) (Cap). A device that, by insertion in a receptacle, establishes connection between the conductors of the attached flexible cord and the conductors connected permanently to the receptacle. (NFPA 70, *National Electrical Code*, 1990 edition.)

Bonding. The permanent joining of metallic parts to form an electrically conductive path that will assure electrical continuity and the capacity to conduct safely any current likely to be imposed. (NFPA 70, *National Electrical Code*, 1990 edition.)

Branch Circuit. The circuit conductors between the final overcurrent device protecting the circuit and the outlet(s). (NFPA 70, *National Electrical Code*, 1990 edition.)

Circuit Breaker. A device designed to open and close a circuit by nonautomatic means and to open the circuit automatically on a predetermined overcurrent without injury to itself when properly applied within its rating. (NFPA 70, *National Electrical Code*, 1990 edition.)

Circuit Interrupter. A nonautomatic operated device designed to open (under abnormal conditions) a current-carrying circuit without injury to itself.

Conduit.

Rigid Metal Conduit. A raceway specially constructed for the purpose of the pulling in or the withdrawing of wires or cables after the conduit is in place and made of metal pipe of standard weight and thickness permitting the cutting of standard threads. (ANSI/IEEE Standard No. 100-1984.)

Intermediate Metal Conduit. A metal raceway of circular cross section with integral or associated couplings, connectors, and fittings approved for the installation of electrical conductors. (NFPA 70, *National Electrical Code*, 1990 edition.)

Rigid Nonmetallic Conduit. A type of conduit and fittings of suitable nonmetallic material that is resistant to moisture and chemical atmospheres, flame retardant, resistant to impact and crushing, and resistant to distortion from heat or low temperatures under conditions likely to be encountered in service. (NFPA 70, *National Electrical Code*, 1990 edition.)

Control Circuit. The circuit of a control apparatus or system that carries the electric signals directing the performance of the controller but does not carry the main power current. (NFPA 70, *National Electrical Code*, 1990 edition.)

Control Circuit Transformer. A voltage transformer utilized to supply a voltage suitable for the operation of control devices. (ANSI/IEEE Standard No. 100-1984.)

Control Circuit Voltage. The voltage provided for the operation of shunt coil magnetic devices.

Control Compartment. A space within the base, frame, or column of the machine used for mounting the control panel. (ANSI/IEEE Standard No. 100-1984.)

Control Enclosure. The housing for the control panel, whether mounted on the machine or separately mounted.

Controlled Stop. The stopping of machine motion by reducing the command signal to zero (0) but retaining power to the machine actuators during the stopping process.

Controller. A device or group of devices that serves to govern, in some predetermined manner, the electric power delivered to the apparatus to which it is connected. For the purpose of this standard, a controller is any switch or device normally used to start and stop a motor by making and breaking the motor circuit current. (NFPA 70, *National Electrical Code*, 1990 edition.)

Device. A unit of an electrical system that is intended to carry but not utilize electric energy. (NFPA 70, *National Electrical Code*, 1990 edition.)

Disconnecting Means. A device, or group of devices, or other means by which the conductors of a circuit can be disconnected from their source of supply. (NFPA 70, *National Electrical Code*, 1990 edition.)

Dry Location. A location not normally subject to dampness or wetness. A location classified as dry may be temporarily subject to dampness or wetness, as in the case of a building under construction. (NFPA 70, *National Electrical Code*, 1990 edition.)

Dwelling Unit. One or more rooms for the use of one or more persons as a housekeeping unit with space for eating, living, and sleeping, and permanent provisions for cooking and sanitation. (NFPA 70, *National Electrical Code*, 1990 edition.)

Electric Controller. A device or group of devices that serves to govern, in some predetermined manner, the electric power delivered to the apparatus to which it is connected. (ANSI/IEEE Standard No. 100-1984.)

Electromechanical. Applied to any device in which electrical energy is used to magnetically cause mechanical movement.

Electronic Control. That part of the electrical equipment containing circuitry whereby conduction of electrons takes place through a vacuum, gas, or semiconductor substance.

Equipment. Includes material, fittings, devices, appliances, fixtures, apparatus, and the like used as a part of, or in connection with, an electrical installation. (NFPA 70, *National Electrical Code*, 1990 edition.)

Exposed (As Applied to Live Parts). Capable of being inadvertently touched or approached nearer than a safe distance by a person. It is applied to parts not suitably guarded, isolated, or insulated. (NFPA 70, *National Electrical Code*, 1990 edition.)

Feeder. All circuit conductors between the service equipment or the source of a separately derived system and the final branch-circuit overcurrent device. (NFPA 70, *National Electrical Code*, 1990 edition.)

Flame Retardant. So constructed or treated that it will not support or convey flame. (ANSI/IEEE Standard No. 100-1984.)

Grounded. Connected to earth or to some conducting body that serves in place of the earth. (NFPA 70, *National Electrical Code*, 1990 edition.)

Grounded Conductor. A system or circuit conductor that is intentionally grounded. (NFPA 70, *National Electrical Code*, 1990 edition.)

Grounding Conductor. A conductor used to connect equipment or the grounded circuit of a wiring system to a grounding electrode or electrodes. (NFPA 70, *National Electrical Code*, 1990 edition.)

Grounding Conductor, Equipment. The conductor used to connect the noncurrent-carrying metal parts of equipment, raceways, and other enclosures to the system grounded conductor, the grounding electrode conductor, or both, at the service equipment or at the source of a separately derived system. (NFPA 70, *National Electrical Code*, 1990 edition.)

Grounding Electrode Conductor. The conductor used to connect the grounding electrode to the equipment grounding conductor, to the grounded conductor, or both, of the circuit at the service equipment or at the source of a separately derived system. (NFPA 70, *National Electrical Code*, 1990 edition.)

Identified (As Applied to Equipment). Recognizable as suitable for the specific purpose, function, use, environment, application, etc., where described in a particular code requirement. (*See "Equipment."*)

NOTE: Suitability of equipment for a specific purpose, environment, or application may be determined by a qualified testing laboratory, inspection agency, or other organization concerned with product evaluation. Such identification may include labeling or listing. (NFPA 70, *National Electrical Code*, 1990 edition.)

In Sight From, Within Sight From, Within Sight. Where this standard specifies that one equipment shall be "in sight from," "within sight from," or "within sight," etc., of another equipment, one of the equipments specified is to be visible and not more than 50 ft (15.24 m) distant from the other. (NFPA 70, *National Electrical Code*, 1990 edition.)

Inrush Current (Solenoid). The inrush current of a solenoid is the steady-state current taken from the line at rated voltage and frequency with the plunger blocked in the rated maximum open position.

Interconnected Conductors. Refers to those connections between subassemblies, panels, chassis, and remotely mounted devices and does not necessarily apply to internal connections of these units.

Intermittent Duty. Operation for alternate intervals of (1) load and no load; or (2) load and rest; or (3) load, no load, and rest. (NFPA 70, *National Electrical Code*, 1990 edition.)

Interrupting Capacity. Interrupting capacity is the highest current at rated voltage that the device can interrupt.

Jogging (Inching). The quickly repeated closure of the circuit to start a motor from rest for the purpose of accomplishing small movements of the driven machine. (ANSI/IEEE Standard No. 100-1984.)

Locked-Rotor Current. The steady-state current taken from the line with the rotor locked and with rated voltage (and rated frequency in the case of alternating-current motors) applied to the motor. (ANSI/IEEE Standard No. 100-1984.)

Moisture Resistant. So constructed or treated that exposure to a moist atmosphere will not readily cause damage. (ANSI/IEEE Standard No. 100-1984.)

Motor-Circuit Switch. A switch intended for use in a motor branch circuit.

NOTE: It is rated in horsepower, and it is capable of interrupting the maximum operating overload current of a motor of the same rating at the rated voltage. (ANSI/IEEE Standard No. 100-1984.)

Operating Overload. The overcurrent to which electric apparatus is subjected during the normal operating conditions that it may encounter.

NOTE 1: The maximum operating overload is considered to be six times normal full-load current for alternating current industrial motors and control apparatus; four times normal full-load current for direct-current industrial motors and control apparatus used for reduced-voltage starting; and ten times normal full-load current for direct-current industrial motors and control apparatus used for full-voltage starting.

NOTE 2: It should be understood that these overloads are currents that may persist for a very short time only, usually a matter of seconds. (ANSI/IEEE Standard No. 100-1984.)

Overcurrent. Any current in excess of the rated current of the equipment or the rated ampacity of the conductor. It may result from overload, short circuit, or ground fault. (NFPA 70, *National Electrical Code*, 1990 edition.)

Overload. Operation of equipment in excess of normal, full-load rating, or of a conductor in excess of rated ampacity that, when it persists for a sufficient length of time, would cause damage or dangerous overheating. A fault, such as a short circuit or a ground fault, is not an overload. (*See "Overcurrent."*) (NFPA 70, *National Electrical Code*, 1990 edition.)

Panel. An element of an electric controller consisting of a slab or plate on which various component parts of the controller are mounted and wired. (ANSI/IEEE Standard No. 100-1984.)

Power Wiring. The circuit used for supplying power from the supply network to equipment or components used for the productive operation.

NOTE: A motor is one example of equipment or component used for productive operation.

Precision Device. A device that will operate within prescribed limits and will consistently repeat operations within those limits.

Programmable Electronic System (PES). A system based on one or more central processing units (CPUs), connected to sensors or actuators, or both, for the purpose of control or monitoring.

NOTE: The term PES includes all elements in the system extending from sensors to other input devices via data highways or other communication paths to the actuators or other output devices.

Qualified Person. One familiar with the construction and operation of the equipment and the hazards involved. (NFPA 70, *National Electrical Code*, 1990 edition.)

Raceway. An enclosed channel designed expressly for holding wires, cables, or busbars, with additional functions as permitted in this standard.

NOTE: Raceways may be of metal or insulating material, and the term includes rigid metal conduit, rigid nonmetallic conduit, intermediate metal conduit, liquidtight flexible metal conduit, liquidtight flexible nonmetallic conduit, flexible metallic tubing, flexible metal conduit, electrical nonmetallic tubing, electrical metallic tubing, underfloor raceways, cellular concrete floor raceways, cellular metal floor raceways, surface raceways, wireways, and busways. (NFPA 70, *National Electrical Code*, 1990 edition.)

Receptacle. A contact device installed at the outlet for the connection of a single attachment plug. (NFPA 70, *National Electrical Code*, 1990 edition.)

Relay. An electric component designed to interpret input conditions in a prescribed manner and, after specified conditions are met, to respond to cause contact operation or similar abrupt change in associated electric control circuits.

NOTE 1: Inputs are usually electric but may be mechanical, thermal, or other types. Limit switches and similar devices are not relays.

NOTE 2: A relay consists of several units, each responsive to specified inputs, the combination providing the desired performance characteristic.

Short-Time Rating. The rating defining the load that can be carried for a short and definitely specified time, with the machine, apparatus, or device being at approximately room temperature at the time the load is applied.

Splashproof (Industrial Control). So constructed and protected that external splashing will not interfere with successful operation. (ANSI/IEEE Standard No. 100-1984.)

Static Device. A device that has no moving parts, as associated with electronic and other control or information-handling circuits.

Subpanel. An assembly of electrical devices connected together that forms a simple functional unit in itself.

Tight (Suffix). So constructed that the specified material is excluded under specified conditions. (ANSI/IEEE Standard No. 100-1984.)

Undervoltage Protection. The effect of a device that operates on the reduction or failure of voltage to cause and maintain the interruption of power.

NOTE: The principal objective of this device is to prevent automatic restarting of the equipment. Standard undervoltage or low-voltage protection devices are not designed to become effective at any specific degree of voltage reduction.

Ventilated. Provided with a means to permit circulation of air sufficient to remove excess heat, fumes, or vapors. (NFPA 70, *National Electrical Code*, 1990 edition.)

Wet Location. Installations underground or in concrete slabs or masonry in direct contact with the earth, and locations subject to saturation with water or other liquids, such as vehicle washing areas, and locations exposed to weather and unprotected. (NFPA 70, *National Electrical Code*, 1990 edition.)

Wireway. A sheet-metal trough with hinged or removable covers for housing and protecting electric wires and cable and in which conductors are laid in place after the wireway has been installed as a complete system. (NFPA 70, *National Electrical Code*, 1990 edition.)

Appendix B Examples of Industrial Machinery Covered by NFPA 79

This Appendix is not a part of the requirements of this NFPA document, but is included for information purposes only.

B-1 Machine Tools

1. Metal Cutting
2. Metal Forming

B-2 Plastics Machinery

1. Injection Molding Machines
2. Extrusion Machinery
3. Blow Molding Machines
4. Specialized Processing Machines
5. Thermoset Molding Machines
6. Size Reduction Equipment

B-3 Wood Machinery

1. Woodworking Machinery
2. Laminating Machinery
3. Sawmill Machines

B-4 Assembly Machines

B-5 Material Handling Machines

1. Industrial Robots
2. Transfer Machines

B-6 Inspection/Testing Machines

1. Coordinate Measuring Machines
2. In-Process Gaging Machines

Appendix C   Graphical Symbols

This Appendix is not a part of the requirements of this NFPA document, but is included for information purposes only.

C-1 Warning Sign on Enclosures
(Based on 417-IEC-5036)

C-2 Equipment Grounding Terminal
(417-IEC-5019)

C-3 Grounded Terminal
(417-IEC-5017)

C-4 Start or On
(417-IEC-5007)

C-5 Stop or Off
(417-IEC-5008)

C-6 Alternatively Act as Start and Stop or On and Off
(417-IEC-5010)

C-7 Movement When Pressed and Stop Movement When Released (Jogging)
(417-IEC-5011)

Appendix D   Sample Electrical Diagrams

This Appendix is not a part of the requirements of this NFPA document, but is included for information purposes only.

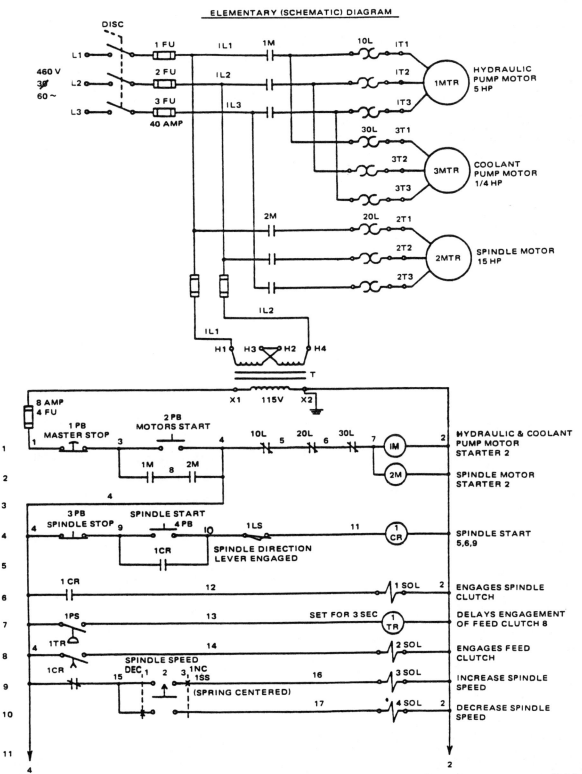

This sample diagram is based on an electromechanical relay logic system. The present state of the art utilizes programmable logic ladder diagrams or computer-generated logic ladder diagrams.

Figure D-1.

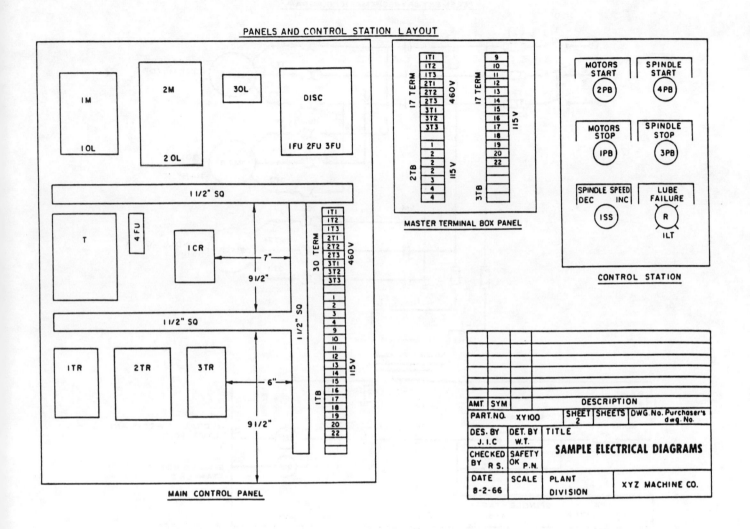

Figure D-2.

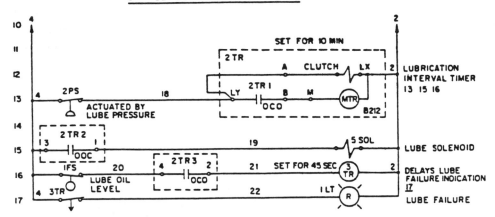

SAMPLE ELECTRICAL DIAGRAMS

SEQUENCE OF OPERATION

A. **Machine Operation:** press "MOTORS START" pushbutton "2PB". Motors Start.
B. Select spindle speed by turning selector switch "1SS" to "INC", energizing "3 SOL", to increase or to "DEC", energizing "4 SOL", to decrease setting.
C. With correct spindle direction selected, limit switch "1LS" is actuated. Press "SPINDLE START" pushbutton "4PB" energizing relay "1CR" which energizes "1 SOL". Spindle starts and pressure switch "1PS" is actuated. "1PS" energizes "1TR" and after a time delay "2 SOL" is energized permitting movement of machine elements at selected feed rates.
D. Pressing "SPINDLE STOP" pushbutton "3PB" stops spindle and feeds movements simultaneously.
E. **Lubrication Operation:**
F. Pressure Switch "2PS" Is Closed.
   1. Timer "2TR" clutch is energized when motors start.
   2. Contact "2TR-1" closes and energizes timer motor "MTR" starting lube timing period.
   3. Contact "2TR-3" closes and energizes timer "3TR".
G. Timer "2TR" Times Out.
   1. Contact "2TR-1" opens, deenergizing timer motor "MTR".
   2. Contact "2TR-2" closes, energizing "5 SOL".
   3. Contact "2TR-3" opens, deenergizing timer "3TR".
   4. Lubrication pressure actuates pressure switch "2PS", de-energizing and resetting timer "2TR". Contacts "2TR-1, 2TR-2, and "2TR-3" open.
   5. Contact "2TR-2" opening, deenergizes "5 SOL".
H. Reduced lubrication pressure deactuates pressure switch "2PS" and sequence repeats.

SWITCH OPERATION

1LS (4) Actuated by spindle direction lever engaged
1PS (11) Operated when spindle clutch engaged
2PS (13) Operated by normal lube pressure
1FS (16) Operated by adequate lube supply
For panels and control station layout see Sheet 2
For hydraulic diagram see _____
For lubrication diagram see _____
Last wire number used 22
Last relay number used 1CR

Supplier's dwg. no. _____
Supplier's name _____
Purchase order no. P.O. 91011
Serial no. of machine TYP 121314

These diagrams used for machine no. _____

AMT	SYM		DESCRIPTION		
			FULL CATALOG DESCRIPTION OF ALL ITEMS		
PART.NO.	XY 100		SHEET 2	SHEETS	DWG.No. Purchaser's dwg.No.
DES.BY J.I.C	DET.BY W.T.		SAMPLE ELECTRICAL DIAGRAMS		
CHECKED BY R.S.	SAFETY OK P.N.				
DATE 8-2-66	SCALE		PLANT DIVISION	XYZ MACHINE CO.	

Figure D-3.

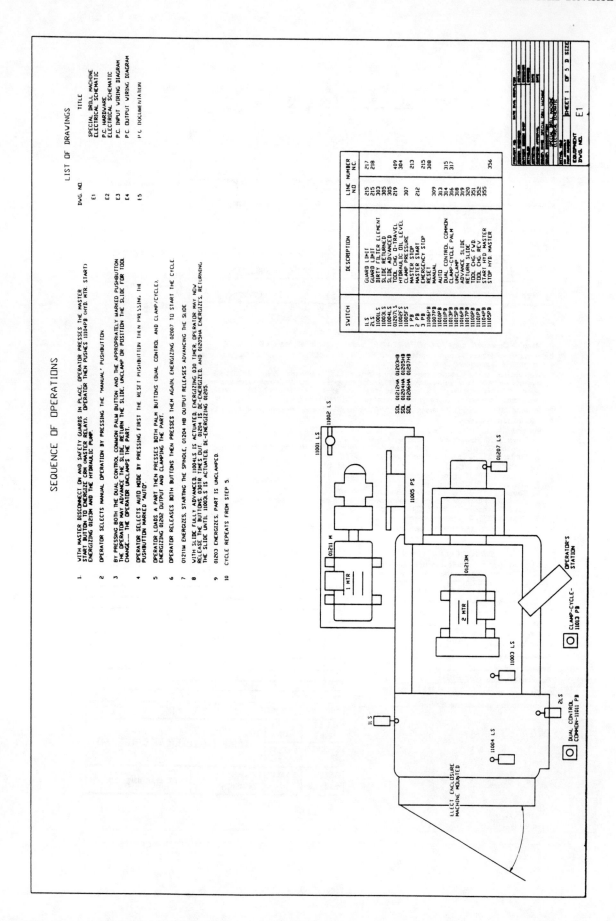

Figure D-4.

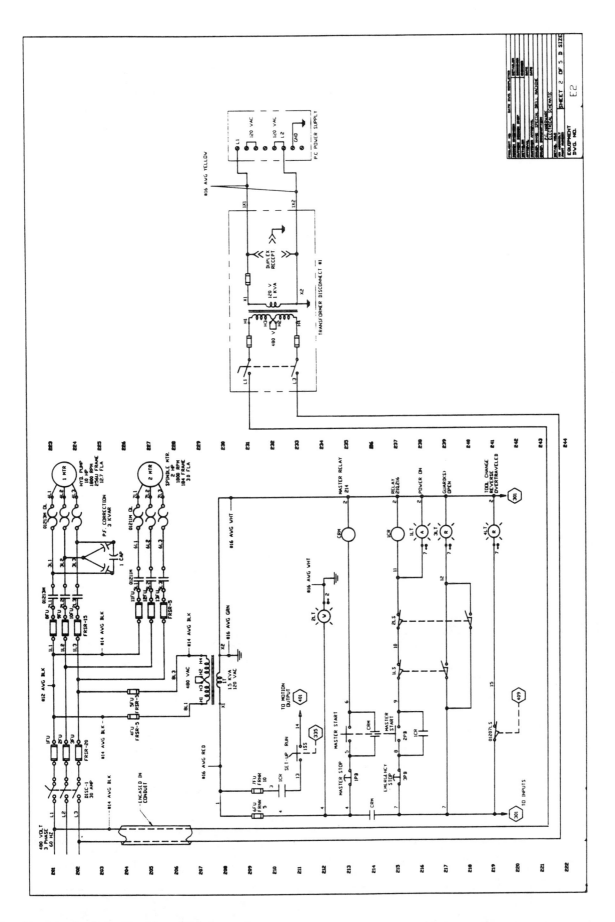

Figure D-5.

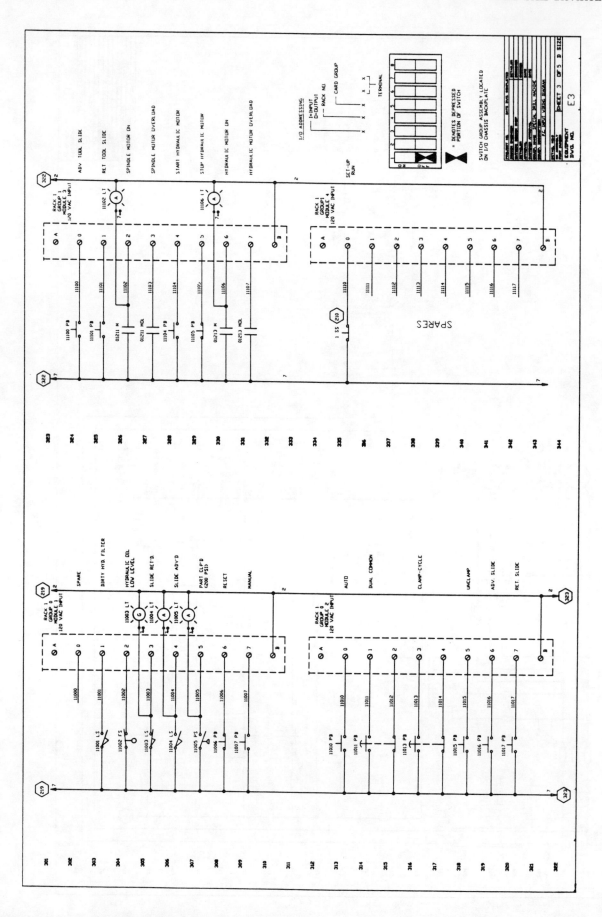

Figure D-6.

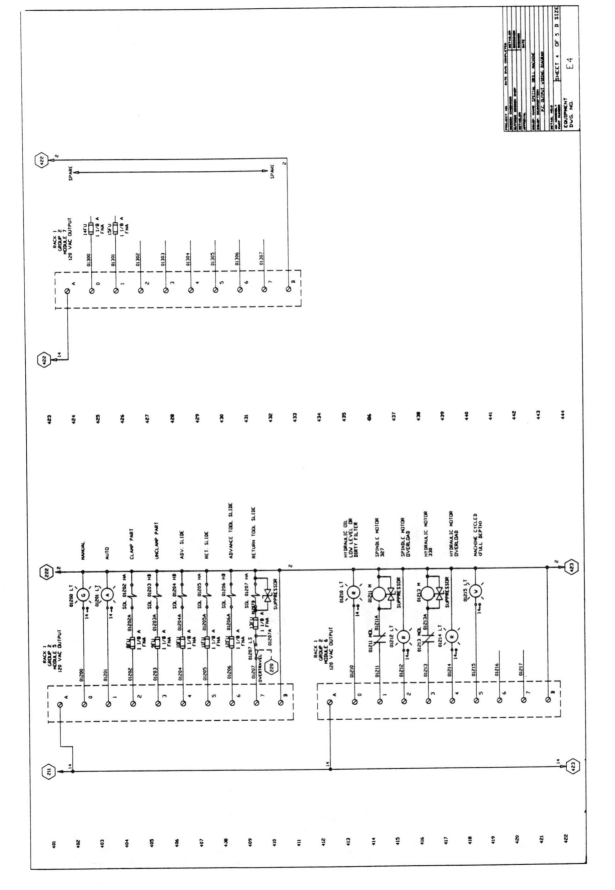

Figure D-7.

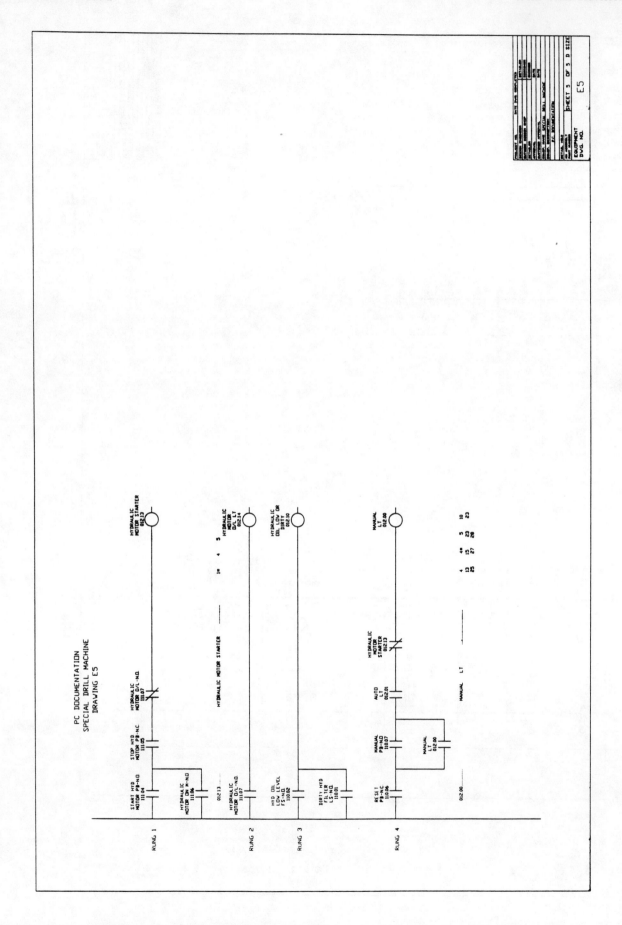

Figure D-8.

Appendix E   Device and Component Designations

This Appendix is not a part of the requirements of this NFPA document, but is included for information purposes only.

The device and component designations given below are intended for use on diagrams in connection with the corresponding graphical symbols to indicate the function of the particular device. These device and component designations are based on the assignment of a standard letter or letters to the fundamental function that is performed by a component or device. Suitable numbers (1, 2, 3, etc.) and letters (A, B, C, etc.) may be added to the basic designation to differentiate between devices performing similar functions.

The assignment of a designation to a device on specific equipment is governed by the function of that device on that equipment and not by the type or nature of the device or its possible use for other functions on other equipment. The same type of device may perform different functions on different equipment or even on the same equipment, and consequently, may be identified by different designations.

Designation	Device
ABE	Alarm or Annunciator Bell
ABU	Alarm or Annunciator Buzzer
AH	Alarm or Annunciator Horn
AM	Ammeter
AMP	Amplifier
AT	Autotransformer
CAP	Capacitor
CB	Circuit Breaker
CI	Circuit Interrupter
CNC	Computerized Numerical Controller
CON	Contactor
COS	Cable Operated (Emergency) Switch
CPU	Central Processing Unit
CR	Control Relay
CRA	Control Relay, Automatic
CRH	Control Relay, Manual
CRL	Control Relay, Latch
CRM	Control Relay, Master
CRT	Cathode Ray Tube, Monitor or Video Display Unit
CRU	Control Relay, Unlatch
CS	Cam Switch
CT	Current Transformer
CTR	Counter
D	Diode
DISC	Disconnect Switch
DISP	Display
DR	Drive
END	Encoder
FLD	Field
FLS	Flow Switch
FS	Float Switch
FTS	Foot Switch
FU	Fuse
GEN	Generator
GRD, GND	Ground

Designation	Device
HM	Hour Meter
HTR	Heating Element
IC	Integrated Circuit
INST	Instrument
IOL	Instantaneous Overload
I/O	Input/Output Device
L	Inductor
LED	Light Emitting Diode
LS	Limit Switch
LT	Pilot Light
LVDT	Linear Variable Differential Transformer
M	Motor Starter
MD	Motion Detector
MF	Motor Starter—Forward
MG	Motor—Generator
MR	Motor Starter—Reverse
MTR	Motor
OL	Overload Relay
PB	Pushbutton
PBL	Pushbutton, Illuminated
PC	Personal Computer
PCB	Printed Circuit Board
PEC	Photoelectric Device
PL	Plug
PLC	Programmable Logic Controller
POT	Potentiometer
PRS	Proximity Switch
PS	Pressure Switch
PWS	Power Supply
Q	Transistor
QTM	Thermistor
REC	Rectifier
RECP	Receptacle
RES	Resistor
RH	Rheostat
S	Switch
SCR	Silicon Controlled Rectifier
SOL	Solenoid
SS	Selector Switch
SSR	Solid State Relay
ST	Saturable Transformer
SUP	Suppressor
SYN	Synchro or Resolver
T	Transformer
TACH	Tachometer Generator
TAS	Temperature Actuated Switch
TB	Terminal Block
T/C	Thermocouple
TR	Timer Relay
TWS	Thumbwheel Switch
V	Electronic Tube
VAR	Varistor
VM	Voltmeter
VR	Voltage Regulator
VS	Vacuum Switch
WLT	Worklight
WM	Wattmeter
X	Reactor
ZSS	Zero Speed Switch

Appendix F   Referenced Publications

F-1 The following documents or portions thereof are referenced within this standard for informational purposes only and thus should not be considered part of the requirements of this document. The edition indicated for each reference is the current edition as of the date of the NFPA issuance of this document.

F-1.1 NFPA Publications.   National Fire Protection Association, 1 Batterymarch Park, P.O. Box 9101, Quincy, MA 02269-9101.

NFPA 70, *National Electrical Code*, 1990 edition

NFPA 70E, *Standard for Electrical Safety Requirements for Employee Workplaces*, 1988 edition

F-1.2 Other Publications.

F-1.2.1 UL Publication.   Underwriters Laboratories Inc., 1285 Walt Whitman Road, Melville, Long Island, NY 11746; 1655 Scott Blvd., Santa Clara, CA 95050; and 333 Pfingsten Rd., Northbrook, IL 60062.

UL 1063-1981, *Standard for Safety, Machine Tool Wires and Cables*

F-1.2.2 ANSI Publications.   American National Standards Institute, 1430 Broadway, New York, NY 10018.

ANSI B11.1-1982, *Mechanical Power Presses*

ANSI B11.2-1982, *Hydraulic Presses*

ANSI B11.3-1982, *Power Press Brakes*

ANSI B11.4-1983, *Shears*

ANSI B11.5-1975 (R1981), *Iron Workers*

ANSI B11.6-1984, *Lathes*

ANSI B11.7-1985, *Cold Headers and Cold Former*

ANSI B11.7a (Supplement)

ANSI B11.8-1983, *Drilling, Milling and Boring*

ANSI B11.9-1975, *Grinding Machines*

ANSI B11.10-1983, *Sawing Machines*

ANSI B11.11-1985, *Gear Cutting Machines*

ANSI B11.12-1983, *Roll Forming and Roll Bending*

ANSI B11.13-1983, *Automatic Screw/Bar and Chucking*

ANSI B11.14-1983, *Coil Slitting Machines*

ANSI B11.15-1984, *Pipe, Tube, and Shape Bending Machines*

ANSI B11.17-1983, *Horizontal Hydraulic Extrusion Presses*

ANSI B11.18-1985, *Coil Processing Systems*

F-1.2.3 ASTM Publications.   American Society for Testing and Materials, 1916 Race Street, Philadelphia, PA 19103.

B 8-86, *Standard for Concentric-Lay-Stranded Copper Conductors, Hard, Medium-Hard, or Soft*

B 33-81 (R-1985), *Tinned Soft or Annealed Copper Wire for Electrical Purposes*

B 174-71 (R-1985), *Specification for Bunch-Stranded Copper Conductors for Electrical Conductors*

B 286-89, *Copper Conductors for Use in Hookup Wire for Electronic Equipment*

F-1.2.4 EIA Publications.   Electronic Industries Association, 2001 Eye Street, NW, Washington, DC 20006.

EIA RS-281-B (1979), *Electrical Construction Standards for Numerical Machine Control*

EIA RS-431 (1976), *Electrical Interface Between Numerical Controls and Machine Tools*

F-1.2.5 IEEE Publications.   Institute of Electrical and Electronic Engineers, Inc., 345 E. 47th Street, New York, NY 10017.

IEEE STD 91-1983, *Graphical Symbols for Logic Functions*

IEEE STD 100-1984, *Standard Dictionary of Electrical and Electronics Terms*

F-1.2.6 IEC Publications.   International Electrotechnical Commission, 1, Rue de Varembre, Geneva, Switzerland (SUISSE).

IEC 204-1-1981 (2nd Edition), *Electrical Equipment of Industrial Machines*

IEC 204-2-1983 (2nd Edition), *Appendices D and E of Publication 204-1-1981*

IEC 417 G-1985, *Graphical Symbols for Use on Equipment*, 7th Supplement.

IEC 550-1977, *Interface Between Numerical Control and Industrial Machines*

(ANSI distributes IEC publications in U.S.)

F-1.2.7 NEMA Publications.   National Electrical Manufacturers Association, 2101 L Street NW, Washington, DC 20037.

NEMA ICS-1-88, *Standards for Industrial Control and Systems*

NEMA ICS-2-1988, *Standards for Industrial Control Devices, Controllers, and Assemblies*

NEMA MG-1-1978, *Motor and Generator Standards*

Variation _____   Rationale _____

Chapter 2   Diagrams, Instructions, and Nameplates

2-1 General.

Replace Section 2-1(a) 5 with:
   5. Subplate & Operator Control Layouts

For consistency: Subplate shall be used to refer to the mounting surface for electrical components instead of Panel or Control Panel.

Operator Control Layouts are required for a complete documentation package.

2-2 Diagrams.

Replace Section 2-2(c) with:
   (c) Symbols for devices shall be identified by a number or number-letter combination, using designations shown in Appendix E. For example:

Control Relay	*CR
Motor Starter	*M
Limit Switch	*LS

   * Numbers may be assigned in sequential order or using the line reference number where the device is located on the drawing. Similar devices on the same line are identified: XXXA LS, XXXB LS, XXXC LS, etc.

   The functional description for each device shall be shown adjacent to its symbol. Functional descriptions shall be in the present or past tense. Motion terminology shall have verb prior to noun and position or status terminology shall have noun prior to verb. For example: Raise Transfer-Transfer Raised; Advance Transfer-Transfer Advanced. Terms such as Transfer Up shall not be used.

This addition provides an example of a consistent method of identifying devices. The exact form to be used must be confirmed with the owner's representatives prior to preparation and submittal of documentation.

Add Section 2-2(j):
   (j) The electrical layout shall consist of a plan view of the equipment showing the control panel enclosure, operator's console, pushbutton station and accessory units not attached directly to the equipment, such as hydraulic power units, in their relative locations. The subplate layout shall show the general physical arrangement of all components on the control panel drawn to scale including identified spare space requirements. Devices may be represented by rectangles or squares and shall be identified as shown on the schematic diagram.

The requirements for detail drawings of the overall plan view of the equipment and subplate layouts are required to insure adequate space for access, maintenance, lock-out tagout locations, and future available spare space.

Variation	Rationale

Add Section 2-2(k):

(k) Circuit schematics and physical layout diagrams shall be provided for electronic assemblies and boards which are user serviceable. The circuit schematics shall provide sufficient component identification and information to allow troubleshooting at the component level. All components shall be identified and cross-referenced with respect to the circuit schematic diagrams.

Added to provide better definition of what documentation is expected from the vendor.

2-5 Warning Marking.

Add Section 2-5(d):

(d) Refer to Chapter 9-10 warning signs on enclosures.

To call attention to a modified warning sign requirement in a new SAE 9-10

2-6 Machine Marking.

Add to Section 2-6:

2-6 Machine Marking. The equipment builder shall provide each equipment with a maintenance safety placard located on the front of the main panel, or similar location of primary maintenance access, diagrammatically identifying the locations of primary energy sources to the machine (electrical, hydraulic, and pneumatic) and the locations of stored energy sources (e.g., capacitors, air tanks, springs, kinetic, gravitational, thermal, and chemical) to be locked out and tagged out per OSHA and NFPA 70E requirements. The placard shall include the lock-out/tag-out method, sequence and confirmation test points to enable maintenance personnel to safely achieve minimal energy state of the machine.

This requirement is intended to insure that the documentation of the lock-out and tagout procedure is included in the original design process and to guarantee that the information is available at the time of installation.

Variation	Rationale

Chapter 3   General Operating Conditions

3-2 Electrical Components and Devices.

Add to Section 3-2:
See Appendix G for the Equipment Data Form to be used by the specifying authority to define special conditions required for the electrical components and devices.

Provides a standard method for specifying authority to provide information specific to their operation.

3-3 Ambient Operating Temperature.

Add to Section 3-3:
The ambient temperature is defined as the temperature surrounding external electrical equipment or control enclosures. Temperature heat rise calculations for internal panel temperatures, and derating factors of components, shall utilize the maximum ambient temperature of 40°C. AMBIENT TEMPERATURE IS NOT INTERNAL ENCLOSURE TEMPERATURE !! Refer to Chapters 8, 13, and 16 for temperature ratings of electrical control equipment and conductors.

Provides a definition for Ambient Temperature and clarification of its use.

3-6 Transportation and Storage.

Add to Section 3-6:
When wireway and conduit are disconnected for shipment, all such openings shall be sealed prior to shipment.

Additional requirement to prevent equipment damage during shipment.

3-7 Voltage Supply

Replace Section 3-7 with the following

The electrical equipment shall be designed to operate correctly under full load as well as no load with the conditions of the nominal supply as specified below unless otherwise specified by the user.

Supply requirements revised to coordinate with proposed NFPA 79 revisions and with IEEE Std. 1100, "Recommended Practices for Powering and Grounding of Sensitive Electronic Equipment.

Table 3-7(a) Voltage Supply AC

(a) Voltage	0.9 ...1.1 of nominal voltage
(b) Frequency	.99...1.01 of nominal frequency continuously. .98...1.02 short time. Note: The short time value may be specified by the user.
(c) Harmonic Distortion	Harmonic distortion not to exceed 10% of the total rms voltage between live conductors for the sum of the 2nd through the 5th harmonic. An additional 2% of the total rms voltage between live conductors for the sum of the 6th through 30th harmonic is permissible.
(d) Voltage unbalance (in 3-phase supply)	Neither the voltage of the negative sequence component nor the voltage of the zero sequence component in 3-phase supplies exceeds 2% of the positive sequence component.
(e) Impulse Voltage	Not to exceed 1.5 ms in duration with a rise/fall time between 500 ns and 500 μs. and a peak value of not more than 200% of the rated rms supply voltage value.
(f) Voltage dips	Short term rms voltage reduced to; 70% of nominal volt. for 30 cycles 42% of nominal volt. for 6 cycles 30% of nominal volt. for 1 cycle There shall be more than 1 sec. between successive dips
(g) Voltage interruptions	Supply disconnected or at zero voltage for 3 ms at any random time in the cycle. More than 1 sec between successive reductions.
(h) Capacitor switching transient	Will operate without disruption with the following transient at maximum rated ac line voltage; 500 Hz ring wave at amplitude equal to rated voltage. (Ref. ANSI Std. C62.45-1992 Sect. B2.3)
(i) Surges / Swells	Short term voltage 30% greater than nominal volt, not to exceed .5 cycles or .0083 sec.
(j) Noise Immunity	Will operate without disruption and within rated performance specifications with the following noise at the ac power terminals; Notch depth - 50% Notch area - 36,000 Volt-microsec. (Ref. IEEE Std. 519-1992 Sect. 10.3)

Table 3-7(b) Voltage Supply DC

From Batteries

(a) Voltage	0.85... 1.15 of nominal voltage; 0.7..1.2 of nominal voltage in case of battery-operated vehicles.
(b) Voltage interruption	Not exceeding 5 ms.

From Converting Equipment

(a) Voltage	0.9... 1.1 of nominal voltage .
(b) Voltage interruption	Not exceeding 20 ms. There shall be more than 1 sec. between successive interruptions.
(c) Ripple (peak to peak)	Does not exceed 0.05 of nominal voltage.

NOTE: Definition of Nominal Voltage: A nominal value assigned to a circuit or system for the purpose of conveniently designating its voltage class (as 120/240, 480Y/277, 600, etc.). The actual voltage can vary from the nominal within a range that permits satisfactory operation of the equipment. See Voltage Ratings for Electrical Power Systems and Equipment (60 Hz), ANSI C84.1-1989. Nominal voltages shall be defined as 120 volts, 208 volts, 240 volts, 480 volts and 600 volts, AC rms.

Variation _____   **Rationale** _____

Chapter 4   Safeguarding of Personnel

No SAE Revisions

Chapter 5   Supply Circuit Disconnecting Means

5-2 Type.

Add to Section 5-2(a):

(1) For all equipment used in systems **greater than 120 volts** the equipment supplier shall furnish one of the following disconnecting means:

   a. Fusible switch conforming with all requirements, except enclosures, listed in NEMA KS-1, and UL 98 for heavy duty type HD switches, rated and tested for 100,000 ampere rms asymmetrical, 600V AC, with Class J, L, R or CC fuses, and with fuse clips of copper alloy, designed to accept the specified Class J, L, R or CC fuses.

   b. A circuit breaker conforming with NEMA AB-1, and UL 489.

   c. A fusible or nonfusible molded-case switch conforming with NEMA AB1 and UL 1087.

The disconnecting means shall have an insulation voltage rating of **600V AC** and service entrance spacing, be horsepower rated, quick-make, quick-break, heavy duty, industrial type and have no exposed live parts in the open position. All current carrying parts shall be of high conductivity copper or copper alloy. All electrical connections shall be metal to metal only. No insulation material shall be used in the pressure system of a joint.

(2) Where nominal **120 volt, single phase, is the only power supply** to the equipment not in accordance with Section 5-11, a fused disconnect switch or circuit breaker of suitable size conforming to the following shall be installed:

   a. Fusible switch conforming with all requirements, except enclosures, listed in NEMA KS-1, and UL 98 for general-duty switches rated **250V AC**, and with fuse clips (where used) of copper alloy, designed to accept the specified Class J, R or CC fuses, or

   b. A circuit breaker conforming with NEMA AB-1, and UL 489.

NOTE: Door interlocking with the disconnect device is not required for enclosures where nominal **120 volt, single phase, is the only power supply** to the equipment. Refer to Section 5-9.

Clearly defines fusible and nonfusible disconnecting means by referencing industry standards.

For additional requirements see section 6-1.

Additional requirement considered necessary for reliability of operation, and safety of equipment in an automotive manufacturing facility

Covers a 120 volt single phase, that is not cord and plug connected. Adds the requirement for a fused disconnect switch or circuit breaker for 120 volt equipment not addressed in 5-2(a.) Also clarifies that door interlocking with the disconnect device is not required.

Variation	Rationale

5-3 Rating.

Replace Section 5-3(b) with:
(b) A disconnecting means with short-circuit and equipment ground-fault protection shall have an interrupting rating sufficient for the nominal circuit voltage and current that is available at the point of installation. (See Equipment Data Form, Appendix G, for plant available short-circuit current.)

A disconnecting means provided to break current at other than fault levels (e.g., motor locked-rotor currents) shall have an interrupting rating that is not less than the largest sum resulting from combining the locked-rotor currents of any combination of motors which can be started simultaneously with the full load currents of the remaining loads that can be operated at that time. This sum is the equivalent locked-rotor current from which the minimum size horsepower rated disconnecting means shall be selected.

Clarifies the requirements of fault level interrupting ratings and non-fault level, e.g., motor locked-rotor currents. Addition required to conform to the current requirements of the NEC Section 110-9.

5-4 Position Indication.

Add to Section 5-4:
The disconnecting means shall have a visible gap or a position indicator which cannot indicate open (off) until all contacts are actually open. See Section 5-10(d) for operating handle requirements.

Clarifies the meaning of "open".

5-5 Supply Conductors to Be Disconnected.

Add Exception to Section 5-5:
Exception: Control devices and terminals located in the enclosure but energized from a separate source need not be de-energized where identified by yellow wiring. The warning requirements of Section 2-4(a) shall apply. See Sections 7-1 and 17-8.

Clarifies apparent conflict between Sections 14-1 and 5-5. Allows the use of yellow wiring for circuits that are not de-energized with the main disconnect switch. Allows the use of power supply taken from the line side of the disconnect to supply memory elements. Refer to Appendix D Figure D-5

Variation   **Rationale**

5-6 Connections to Supply Lines.

Add to Section 5-6:
The incoming supply line conductors shall be contained in their own separate conduit or raceway.

Requires the design of the supply conductors be separated from other conductors entering or leaving the control enclosure to eliminate the possibility of line to load conductor faults.

5-7 Exposed Live Parts.

Add to Section 5-7 (above NOTE):
Where test probe access holes are provided, they shall be insulated to guard against accidental shorting between line terminals and adjacent grounded metal.

Provides for the safe use of test probes.

5-8 Mounting.

Add to Section 5-8(a):
Mounting the disconnecting means "adjacent thereto" is permitted only as described in Exception No. 3 below.

Preferred installation is with the disconnecting means mounted within the main enclosure for interlocking purposes.

Add Exception No. 3 to Section 5-8(a):
Exception No. 3: Where panel mounting of a disconnecting means rated more than 600 amperes is impractical, a disconnecting means in a separate enclosure shall be permitted with approval of the specifying authority.

Add Section 5-8(c):
(c) Where a "main" disconnecting means is required for remote main accessible lockout, adequate clearance shall be provided directly below the main disconnecting means for wiring termination and / or power distribution blocks. The additional disconnecting means shall be provided as described below:

Describes additional requirements for the disconnecting means that are external to the main control enclosure. (not addressed by NFPA 79)

(1) Be mounted in an enclosure located at the operator console or other readily accessible location.

(2) Be connected to the load side of the line disconnecting means in the main control enclosure.

(3) Have the line and load side conductors for each disconnecting means installed in separate raceways.

Variation	Rationale

5-9 Interlocking.

Replace Section 5-9(a) with:
(a) Each disconnecting means shall be interlocked with its control enclosure door(s). Interlocking shall be provided between the disconnecting means and its associated master door to accomplish the following:

(Exception Nos. 1-4 in NFPA 79 Section 5-9(a) remain)

1. Prevent closing of the disconnecting means while the master door is open, unless the disconnect interlock is operated manually.

2. Prevent closing of the disconnect means while the master door is in the initial latch position or until the vault hardware is fully engaged.

3. Prevent opening the master door until the disconnect is opened, or the mechanical interlock is bypassed.

The interlock operation shall allow additional doors to be closed in any order. The master door must be opened before other doors can be opened. When the master door is closed first, the other doors must interlock automatically when they are closed and latched. Progressive interlocking (door-to-door) shall not be used.

In a multi-door enclosure, each door providing access to control components where the maximum voltage is 150 Vac or less, shall be permitted to be opened independently of the master door. Any such doors that are not interlocked with the master door shall include a latch operated by a key or a common hand tool.

Defines how the interlock should function. Also allows noninterlocking doors for applications of 150 volts or less.

5-10 Operating Handle.

Replace Section 5-10(a) with:
(a) The operating handle of the disconnecting means shall be readily accessible with doors in the open or closed position. The mechanical linkage between the disconnecting means and its operating handle shall be such that the operating handle is in control of the disconnect at all times. Operating handles shall maintain the NEMA rating of the enclosure.

Required so that the operating handle is in control of the disconnecting means at all times. Ensures integrity of the enclosure.

Replace Section 5-10(b) with:
(b) The center of the grip of the operating handle of the disconnecting means, when at its highest position, shall be not more than 2 m (6.5 ft) or less than .9m (3 ft) above the servicing level.

The paragraph is reworded to use the defined term, "servicing level" and provides minimum height requirement.

Add to Section 5-10(c):
Provision shall be made for a minimum of three locks having shackles 7.9 mm (.312 in.) in diameter. The design shall be such that the switch cannot be closed when locked in the open (off) position by a single padlock or locking device.

Provides for a multi-shift or multi-trade lock-out tag-out procedure without additional hardware.

<u>Variation</u>   <u>Rationale</u>

Add to Section 5-10(d):
Where color is used, the ON position shall be red. Solid plastics or similar material should be used for colored knobs or handle inserts; paint is not acceptable.

Add Section 5-10(e):
 (e) The open (off) position of the operating handle shall be down for vertical operation or to the right for horizontal operation.

Further defines the handle operation for the off position, which is not noted in NFPA.

Variation	Rationale

Chapter 6   Protection

6-1 Machine Circuits.

Replace Section 6-1 with:
Figures 6-1 and 6-2 show typical circuits acceptable for the protection of current carrying and current consuming electrical machine components. Protective interlocks are not shown.

Figure 6-2 was added to clarify protection of machine tool transformers and drive isolation transformers. The scope of this section has been expanded to broaden its applicability

Add Section 6-1(a):
(a) Branch circuit overcurrent protective devices shall be selected to protect components during the occurrence of short circuits and equipment ground faults. All protective devices shall be selected and applied with proper consideration being given to, but not limited to, the following:

Defines the purpose of Overcurrent protection as described in chapter 6, Sub-sections 1 through 6. Give general criteria used in the selection of protective devices for the balance of Chapter 6.

1. System maximum available fault current at the point of application. *(See equipment data form for available fault current)*

2. Interrupting rating of the protective device.

3. Voltage rating of the system.

4. Load and circuit component characteristics:
   a. Normal operating current
   b. Inrush characteristics

5. Control device rated conditional short-circuit current.

See definition in Appendix A

6. Coordination of protective devices with each other.

Add Section 6-1(b):
(b) Fuses, fuseblocks and fusible switches shall be identified by a recognized testing agency and shall be of the following Classes: L, RK1, RK5, J, CC, Fuses shall also be identified as time delay. CLASS H FUSES (RENEWABLE AND ONE-TIME), FUSEBLOCKS AND FUSIBLE SWITCHES SHALL NOT BE USED IN ANY APPLICATIONS. Special application fuses other than those listed above shall be permitted to be used as a component of a subsystem. Fuses, fuseblocks and fusible switches used in control circuits not over 125 volts, 30 amperes shall be permitted to be supplementary type with a minimum interrupting capacity of 1,000 amperes at 125 volts.

Underwriters Laboratories classifications of fuses, blocks and switches are defined. This eliminates the lower interrupting capacity class H equipment. Time delay fuses are specified for all installations to avoid duplication of maintenance spare parts stocking.

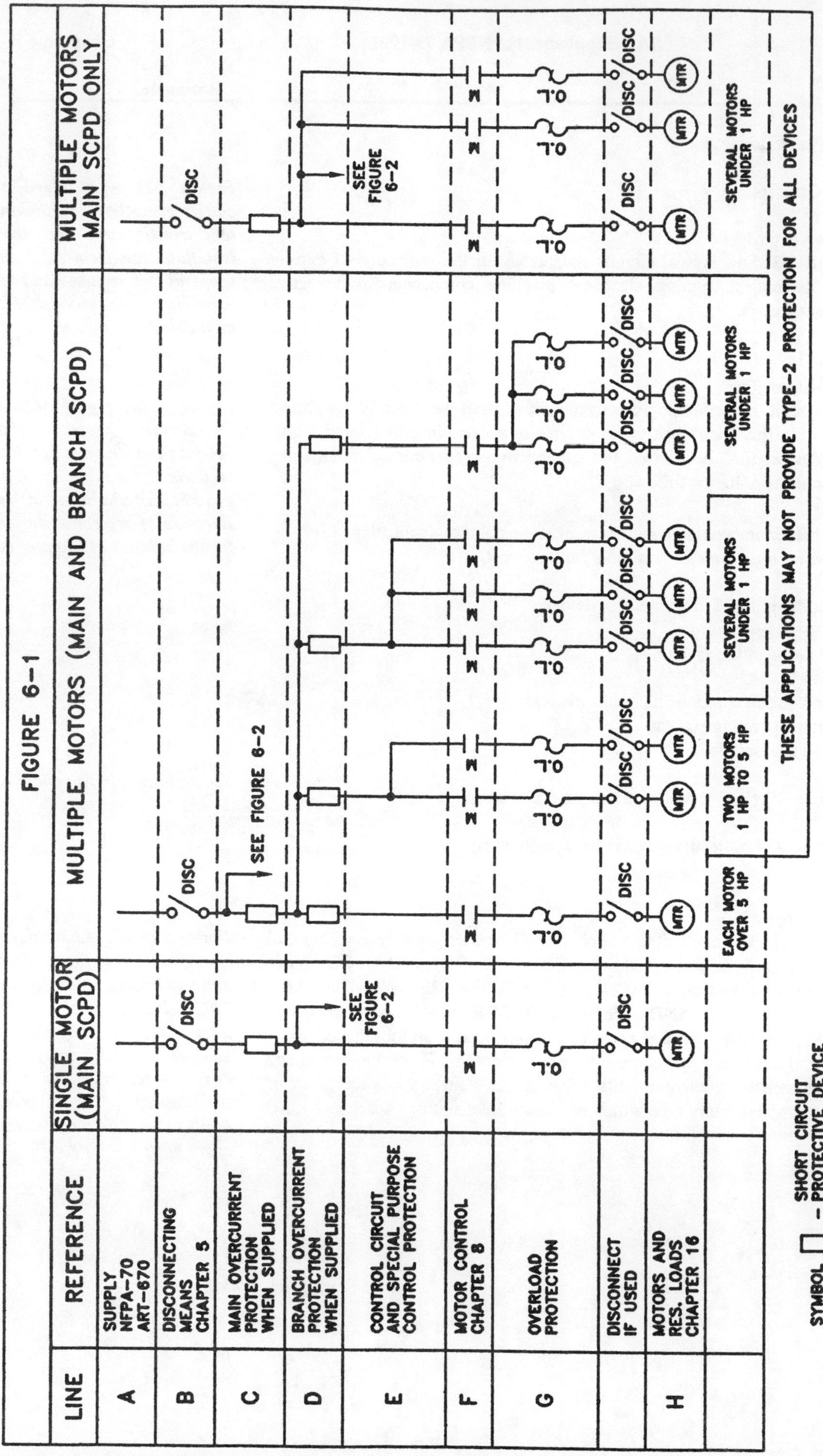

FIGURE 6-1

LINE	REFERENCE	SINGLE MOTOR (MAIN SCPD)	MULTIPLE MOTORS (MAIN AND BRANCH SCPD)	MULTIPLE MOTORS MAIN SCPD ONLY
A	SUPPLY NFPA-70 ART-670			
B	DISCONNECTING MEANS CHAPTER 5	DISC	DISC	DISC
C	MAIN OVERCURRENT PROTECTION WHEN SUPPLIED			
D	BRANCH OVERCURRENT PROTECTION WHEN SUPPLIED	SEE FIGURE 6-2	SEE FIGURE 6-2	SEE FIGURE 6-2
E	CONTROL CIRCUIT AND SPECIAL PURPOSE CONTROL PROTECTION			
F	MOTOR CONTROL CHAPTER 8	M	M	M
G	OVERLOAD PROTECTION	O.L.	O.L.	O.L.
	DISCONNECT IF USED	DISC	DISC	DISC
H	MOTORS AND RES. LOADS CHAPTER 16	MTR	EACH MOTOR OVER 5 HP / TWO MOTORS 1 HP TO 5 HP / SEVERAL MOTORS UNDER 1 HP / SEVERAL MOTORS UNDER 1 HP	SEVERAL MOTORS UNDER 1 HP

THESE APPLICATIONS MAY NOT PROVIDE TYPE-2 PROTECTION FOR ALL DEVICES

SYMBOL ☐ – SHORT CIRCUIT – PROTECTIVE DEVICE (SCPD)

FIGURE 6-1 – ONE LINE REPRESENTATION OF ELECTRICAL SYSTEM POWER PROTECTION

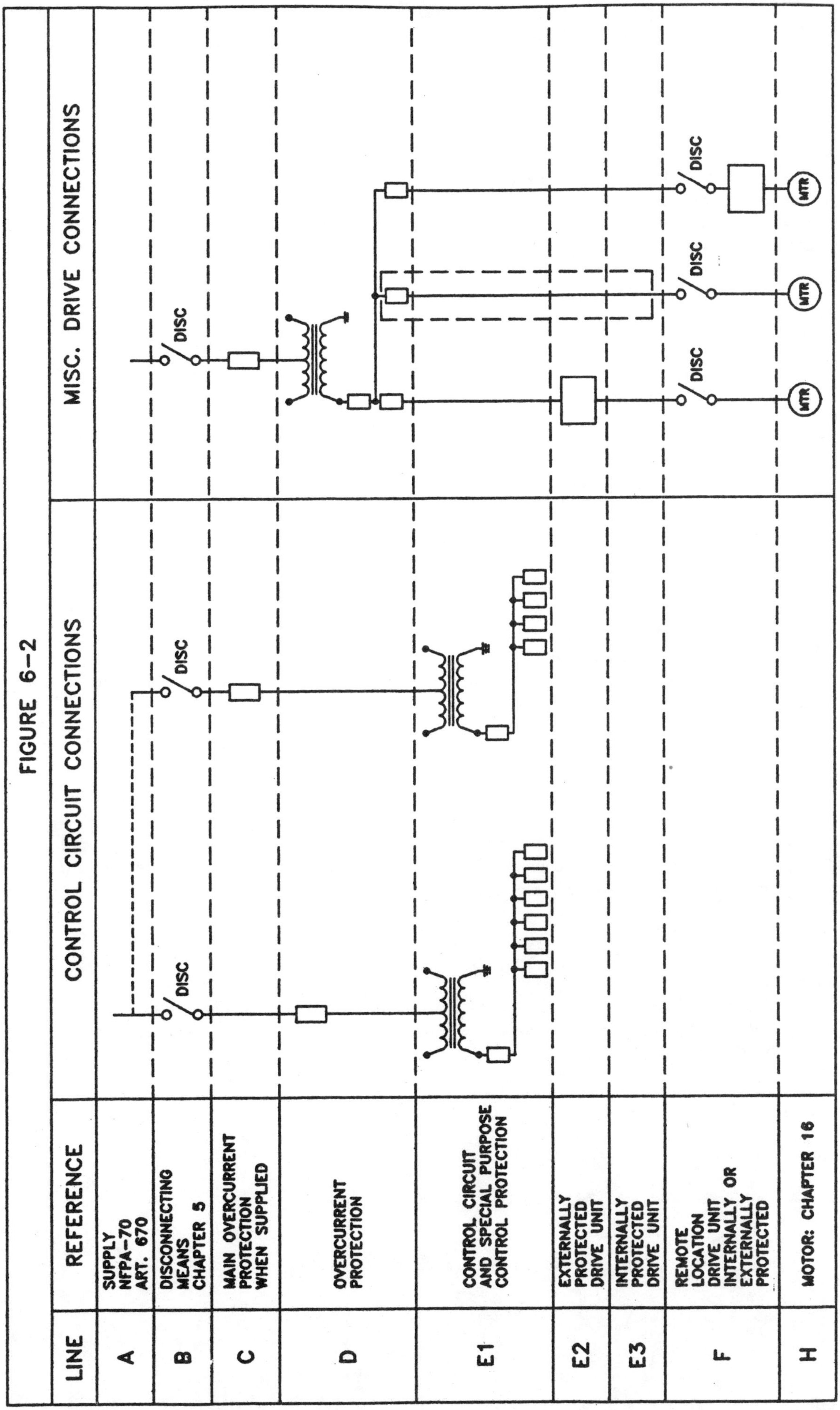

FIGURE 6-2

LINE	REFERENCE	CONTROL CIRCUIT CONNECTIONS	MISC. DRIVE CONNECTIONS
A	SUPPLY NFPA-70 ART. 670		
B	DISCONNECTING MEANS CHAPTER 5		
C	MAIN OVERCURRENT PROTECTION WHEN SUPPLIED		
D	OVERCURRENT PROTECTION		
E1	CONTROL CIRCUIT AND SPECIAL PURPOSE CONTROL PROTECTION		
E2	EXTERNALLY PROTECTED DRIVE UNIT		
E3	INTERNALLY PROTECTED DRIVE UNIT		
F	REMOTE LOCATION DRIVE UNIT INTERNALLY OR EXTERNALLY PROTECTED		
H	MOTOR: CHAPTER 16		

FIGURE 6-2 — ONE LINE REPRESENTATION OF ELECTRICAL SYSTEM POWER PROTECTION

Variation	Rationale

Add Section 6-1(c):

(c) All circuit breakers shall be identified by a recognized testing agency and shall be suitable for the specific application as outlined in 6-1(a). Circuit breakers used in control circuits not over 125 volts, 30 amperes shall have a minimum interrupting capacity of 1,000 amperes.

Same basic intent as 6-3(b).

6-2 Supply Conductor and Machine Overcurrent Protection.

Add Section 6-2(a):

(a) Machine supply conductors shall terminate at a single disconnecting means and set of overcurrent protective devices. The recommended size or rating of the overcurrent protective device shall be a rating or setting not greater than the largest rating or setting of the branch-circuit short-circuit and ground-fault protective device for any motor of the group, plus the sum of the full-load currents of the other motors of the group, plus any other loads. Where the sum of these currents does not match an available size or rating the next larger size or rating shall be used.

Automotive requirements are for each machine to have a main disconnect and overcurrent protection. This is in harmony with NFPA 70 (1993) 430-62(a) and 430-24.

Add Section 6-2(b):

(b) Where two or more protective devices are applied in series they shall be selected with the proper time-current and let-thru energy characteristics to optimize power system selective coordination.   Determination shall be based on the manufacturer's published data.

Automotive productivity demands require this selective coordination of protective devices; as does NFPA 70 (1993) 240-12

6-3 Additional Overcurrent Protection.

Add Section 6-3(a):

(a) Programmable Electronic Systems (PES) power supply inputs shall be protected by an overcurrent protective device either externally or internally. The overcurrent protection size or rating shall be in accordance with the manufacturer's instructions.

Appropriate overcurrent protection is needed in case of a failure within the power supply.

Add Section 6-3(b):

(b) Programmable Electronic Systems (PES) outputs shall be either internally protected or provided with external overcurrent protective devices sized or rated in accordance with the manufacturers instructions.

The output circuitry needs overcurrent protection to minimize damage in case of a fault.

Variation	Rationale

Add Section 6-3(c):

(c) Pushbuttons, Selector Switches, Limit Switches, Solid-state sensors and proximity switches shall in no case be connected to a circuit rated more than 10 amperes.   When control transformers 1 KVA and larger are used the current limiting ability of the overcurrent protective device should be considered.

NOTE:
Also see SAE Inductive and Capacitive Proximity Sensor Specification SAE J1738-1 Section 6.3 Electrical Protection.

UL limits the size of the power supply used during the test procedure.   The use of large control transformers (over 1 KVA) often exceeds the withstand rating of the component.

6-4 Location of Protective Devices.

Retitle Section 6-4 with:
6-4 Location of Protective Devices for Motor Circuits.

The existing NFPA 79 text deals only with motor circuits.   This revision expands the requirements.

Add at the end of Section 6-4

NOTE: For the location of protective devices for nonmotor loads, see NFPA 70 (NEC), 1993 Article 240-21.

The content of this Section addresses the differences outlined in NFPA 70 for motor circuits and nonmotor circuits.

6-5 Motor Branch Circuits

Retitle Section 6-5:
6-5 Motor Branch Circuit and Motor Controller.

Replace Section 6-5(a):

(a) Except for conditions indicated in 6-5(d), each motor controller and its associated wiring, shall be protected as an individual branch-circuit by a Short Circuit Protective Device (SCPD) as specified by the controller manufacturer. The maximum rating of the designated SCPD shall be as shown in table 6-5(a).

UL 508, NFPA 70, and NFPA 79 do not address the subject of motor controller damage protection. IEC 204 and IEC 947-4-1 define the terms type 1 and type 2 coordinated protection. Automotive users recognize the need for Type 2 protection as being an issue of reliability.
These application considerations introduce the machine builder to new and valuable information.

Variation   Rationale

Retitle and Replace Table 6-5(a) with:

Table 6-5(a)   Fuse and Circuit Breaker Selection Motor, Motor Branch-Circuit, and Motor Controller

Maximum Setting or Rating (*1) (Fuse and Circuit Breaker)					
Application	Fuse Class with Time Delay (*3)				Inst. Trip	Inverse Time
Type (*2)	RK-5 (*4)	RK-1	J	CC	C/B (*5)	C/B (*6)
AC-2	150%	150%	150%	150%	700%	150%
AC-3	175%	175%	175%	300%	700%	250%
AC-4	175%	175%	225%	300%	700%	250%

NOTES:

(*1) For motors with locked rotor code letters A thru E, see NFPA 70 1993, Table 430-152.

(*2) TYPES OF APPLICATIONS:
   AC-2 Slip-ring motors: starting, switching off. or All light starting duty motors.
   AC-3 Squirrel-cage motors: starting, switching off while running; occasional inching, jogging or plugging but not to exceed 5 operations per minute or 10 operations per ten minutes. All wye-delta and two step auto-transformer   starting. or All medium starting duty motors.
   AC-4 Squirrel-cage motors: starting, plugging, inching jogging. or All heavy starting duty motors.

(*3) Where the rating of a time delay fuse (other than class CC) is not sufficient to start the motor it shall be permitted to be increased to 225%; class CC fuses shall be permitted to be increased to 400%.

(*4) Class RK-5 fuses shall be used only with NEMA rated motor controllers.

(*5) Magnetic Only Circuit Breakers are limited to single motor applications. These instantaneous trip circuit breakers shall only be used if they are adjustable, part of a combination controller having motor-running and also short-circuit and groundfault protection in each conductor, and if the combination is especially identified for use, and is installed per any instructions included in its listing or labeling. Circuit Breakers with adjustable trip settings shall be set at the Manufacturers' recommendation, but not greater than 1300 percent of the motor full-load current.

(*6) The rating of an inverse time circuit breaker shall be permitted to be increased but in no case exceed (a) 400 percent for full-load currents of 100 amperes or less; or (b) 300 percent for full-load currents greater than 100 amperes.

NOTE:
IEC 947-4-1 defines the terms Type 1 and Type 2 Coordinated Protection as follows:
   Type 1 Protection - Under short circuit conditions the contactor or starter may not be suitable for further use without repair or replacement
   Type 2 Protection - Under short circuit conditions the contactor or starter shall be suitable for further use.

The rating of the designated SCPD shown in table 6-5(a) are the maximum allowed and do not guarantee Type 2 protection. Type 2 Coordinated Protection shall be certified (or witnessed and verified) by a recognized testing facility.

APPLICATION OF TYPE 1 PROTECTION ON SINGLE OR MULTIPLE MOTOR BRANCH CIRCUITS SHALL REQUIRE THE WRITTEN PERMISSION OF THE REQUISITIONING AUTHORITY.

Replace Section 6-5(d):

(d) Multiple motors shall be permitted to be on the same branch circuit if all of the following conditions are met:

   1. No more than two motors from one to five horsepower.

   2. The rating or setting of the overcurrent protective device shall be as low as practicable and shall not exceed the values in Table 6-5(d) for the smallest conductor in the circuit.

   3. The motor controller is identified for use on multiple motor circuits and installed per the listing instructions.

   4. Each motor has its own overload protection as outlined in Section 6-6.

The intent of this change is to restrict the number of motors that can be placed on a single branch circuit protected by a single set of fuses or a circuit breaker.

Variation   **Rationale**

Table 6-5(d)   Relationship Between Conductor Size and Maximum Rating or Short-Circuit Protective Device for Power Circuits

Retitle and replace Table 6-5(d) with:

Table 6-5(d)   Overcurrent Protection for Multiple Motors on a Single Branch Circuit

Conductor Size AWG	Maximum Rating Amperes Dual Element Fuse	Maximum Rating Amperes Inverse Time Circuit Breaker
14	25	35
12	30	45
10	40	70

6-6 Motor Overload.

Add Section 6-6(d):

(d) Each continuous duty motor shall have an overload device that is responsive to motor current. This device shall be selected to trip or be rated at no more than the percent of the full load current shown in Table 6-6(d).

Add Table 6-6(d)   Motor Overload Sizing:

Section 6-6(d) and Table 6-6(d) is from NFPA 70, 1990 section 430-37 and was modified to include a high ambient 1.0 Service Factor motor.

Table 6-6(d)   Motor Overload Sizing

Motors with a marked service factor not less than 1.15	125%
Motors with a marked temperature rise not over 40° C	125%
All other motors	115%
Motors with a 1.0 service factor marked 65° C ambient	125%

Add Section 6-6 (e)

(g) Overload relays shall include a provision for activating an alarm system when the overload sensing device has tripped (operated). A link to a communication system that transmits motor controller status shall be permitted to serve as this tie to the alarm system.

Intent of this section is to specify devices which may have an alarm .

Add Section 6-6 (f) Phase-Loss Protection:

(f) When specified on the Equipment Data Form, phase-loss protection shall be required. Such phase-loss devices shall perform their intended function while other motors in the same electrical distribution network are generating a voltage for the missing phase, and shall disconnect the motor whenever a phase has been lost.

Certain applications require close monitoring of phase-loss.

6-9 Control Circuit Conductors.

Replace Section 6-9(b) with:
  (b) The rating of overcurrent protective devices in the control circuit shall be as low as practicable and shall not exceed the values given in Table 6-9(b) for the smallest conductor in the circuit.

Variation   **Rationale**

Add Table 6-9(b)   **Control Conductor Overcurrent Protection:**

Conductor Size AWG	Maximum Rating Amperes
22	3.5
20	6.25
18	8
16	10
14	15
12	20
10	30
8	40
6	60
4	70
3	80
2	100
1	110
0	125
00	150
000	175
0000	200

NOTE: The overcurrent device rating shall be as low as practical and shall not exceed values specified.

6-11 Power Transformer.

Delete section 6-11 and leave blank intentionally.

Requirements for 6-11 are
included in new section 6-12.

6-12 Control Circuit Transformer.

Retitle and replace Section 6-12 with:
6-12 Control Circuit Transformers, Lighting /Auxiliary Power Transformer
Disconnects, Servo, Drive Isolation, and Specialty Transformers

Replace Section 6-12(a) with:
(a) Each transformer with a primary rated at 600 volts or less shall be protected on the primary and the secondary side by a single overcurrent protective device in each ungrounded conductor to be selected from Table 6-12(a).

| Variation | Rationale |

Add Table 6-12(a)   Overcurrent Protective Device Selection Transformers:

Maximum Full Load Current Rating				
Application	UL Class with Time Delay			Inverse Time
Type	RK5	RK1	J	CC	Circuit Breaker
Primary	250%	250%	250%	250%	200%
Secondary (*1)	125%	125%	125%	125%	125%

Exception; Fuses may be sized at 500% of a transformer primary current at rated load if that current is less than 2 amperes.

NOTE:
(*1) Supplementary Type fuses shall be permitted to be used on the secondary of a control transformers with a maximum deliverable fault current of less than 1,000 amperes.

Replace Section 6-12(b) with:
  (b) Each transformer shall have primary overcurrent protection installed as close as practical to the primary taps of the transformer in accordance with Table 6-12(a).

Primary overcurrent protective devices are intended to isolate faulted transformers from the power distribution system.

Replace Section 6-12(c) with:
  (c) Each transformer shall have secondary overcurrent protection installed as close as practical to the secondary taps of the transformer in accordance with Table 6-12(a).

Secondary overcurrent protection is intended to provide overload protection for the transformer providing a main secondary overcurrent protective device to be sized up to 250% of the rated primary current to reduce undesirable tripping.

Replace Section 6-12(d) with:
  (d) Other overcurrent protection may be required on the transformer secondary branch circuits. Specific load or conductor protection may require the addition of multiple overcurrent protection on the load side of the transformer secondary overcurrent protective device.

6-15 Adjustable-Speed Drive System.

Add Section 6-15(a):
  (a) The incoming branch circuit or feeder shall be protected by an overcurrent protective device in each ungrounded conductor .

It is often confused that adjustable speed drives are a system that is exempt from codes and standards. This section draws attention to protection of the feeder conductor.

To match NFPA 70 (1993) 450-3(b) (Transformers 600 volt and less).

Variation	Rationale

6-18 Power Factor Correction Capacitors.

Add Section 6-18 Factor Correction Capacitors.

Add Section 6-18(a) and (b):
  (a) Where capacitors are installed in circuits, overcurrent protection for the conductors shall be installed, and shall be sized to allow the capacitor to be energized.

Power factor correction capacitor circuits require proper conductor protection. Care should be taken when sizing fuses for conductors that they will not nuisance blow during initial energization.

  (b) Each capacitor (or capacitor bank) shall be protected against rupture of the individual cells per the manufacturers' recommendations. Protection included as a part of the capacitor assembly shall be permitted.

Variation	Rationale

Chapter 7   Control Circuits

7-1 Source of Control Power.

Retitle Section 7-1 as Section 7-1(a).

To add 7-1(b) below.

Replace Section 7-1(a) Exception with:
Exception: An additional supply source may be taken from the line side of the main disconnecting means for programmable devices, auxiliary lighting, receptacles for electronic equipment, and equipment convenience receptacles. This supply source shall comply with the following conditions:

To provide power for other required auxiliary devices and lighting.

1. The line side supply connection is internal to the control panel enclosure and terminates in a totally enclosed transformer disconnecting means as defined in Section 12-3(b)5.

Sources of power other than the main feeder shall not be used to provide control enclosure lighting or control power.

2. The additional supply wiring is separate from and does not share wireway with other conductors.

It is the intent of this exception to allow a lighting transformer disconnect, mounted on the top of a main control enclosure to be connected to the line side of the main disconnecting device.

3. The transformer disconnecting means for this additional supply source is physically mounted adjacent to the main disconnecting means, or the supply wiring is encased in rigid or flexible conduit when the supply conductors are longer than 18 inches. Refer to the tap rules in Chapter 6 for sizing of the supply conductors.

4. The marking requirements of Section 2-5 (a) are complied with.

Add Section 7-1(b):
(b) Where two control power transformers are used, one as a normal control voltage source and one for conditioning (such as supplying filtered and/or regulated power to a programmable device), both transformers shall have their primary windings connected across the same phase.

To preclude phase displacement where normal control source becomes connected to programmable controller I/O modules.

Add Section 7-1(c):
(c) All control power transformers shall have a minimum of 25 percent spare capacity, but not less than 100 volt amperes nor more than 1000 volt amperes. (Also see Section 8-7).

Exception: This requirement does not apply to control power transformers used in combination motor starters.

To permit reasonable expansion of control panels with additional devices requiring control power, without a need to increase the size of the control power transformer.

Add Section 7-1(d):
(d) Any conversion equipment required for systems operation, such as ac to dc, 60 Hz to some other frequency, etc., for any electrical apparatus, shall be furnished by the equipment supplier unless otherwise indicated by the specifying authority.

To insure a complete working system from a single supplier unless the specifying authority is willing to accept system responsibility.

Variation	Rationale

7-2 Control Circuit Voltages.

Add Section 7-2(c):

(c) The common of different voltages and voltage sources shall be isolated from each other. Separate I/O modules, isolated I/O modules, and/or relay contacts shall be used to provide the required isolation.

To prevent unintentional connection of different voltage supplies and unreliable circuit operation.

7-4 Connection of Control Devices.

Delete Section 7-4(a) Exception No. 1.

Has no practical value.

Delete Section 7-4(a) Exception No. 6.

Has no practical justification.

Add to Section 7-4(b):

Not more than ten control circuit contacts shall be connected in series to operate a relay or other coil-operated control device. This limitation includes any combination of limit switch, relay, timer and similar control contacts.

Voltage drop over many contacts cause unreliable operation and too many contacts in series results in circuits which are complicated and difficult to troubleshoot.

Add Section 7-4(c):

(c) Where the use of illuminated pushbuttons is approved by the specifying authority, the following shall apply:

(1) A separately energized LAMP TEST circuit shall be provided to test all lights on the operator's station containing the illuminated pushbuttons. Each operator's station shall have an individual LAMP TEST button.

(2) Each operator's station shall include warning plates or some means of alerting operators to the fact that illuminated pushbuttons are in use and to use the LAMP TEST feature.

(3) Both push-to-test lights and illuminated pushbuttons shall not be installed on the same panel door or control station.

Inherent in the concept of an illuminated pushbutton is the high probability of the device being perceived as a single function device, with confusion being the result of such a perception. Additional features are required to offset this perception.

Add Section 7-4(d):

(d) Transient voltage suppression shall be provided where a contact is hard-wired between a semiconductor output and an inductive load. Where surge suppressors are used to minimize electrical noise, they shall be of the diode, MOV, or RC type and properly rated for the application. Suppressors shall be mounted to eliminate failure of connecting leads due to vibration or exposure to physical damage.

Experience has shown that transient overvoltages are created when a contact breaks an inductive circuit. These overvoltages must be suppressed to prevent damage to semiconductors.

Variation	Rationale

Add Section 7-4(e):

(e) Independent and separate position sensors instead of multiple contacts on a single position sensor shall be used in control circuits that are not de-energized by the same main disconnecting means.

*To provide a proper level of safety by **anticipating** a potential source of misunderstandings when sensors must be disconnected for replacement.*

7-5 Stop Circuits.

Add to Section 7-5(a):
Removing the cause of a stop function shall not restart any part of the equipment.

*An implied **requirement that** should be a stated requirement.*

Add to Section 7-5(b):
Resetting of stop circuitry shall not restart any part of the equipment.

*An implied **requirement that** should be a stated requirement.*

Add to Section 7-5(e):
Equipment that includes a master control relay (see Section 7-19) shall have a CONTROL POWER OFF pushbutton or selector switch that provides another type of stop circuit not shown in table 7-5.

*Many industrial control panels include a master control relay **that provides a** form of stop **function which** should be **included in this** paragraph.*

Add Table 7-5

Table 7-5
Types of Stop Circuits

*Provides an **easy-to-understand** method of **explaining the** different types of stop circuits.*

EMERGENCY STOP	NORMAL STOP	ZONE STOP
1. Shuts down entire machine immediately without creating an additional hazard.	1. Shuts down entire machine in an orderly progression to the end of a cycle.	1. Shuts down a portion of a machine in an orderly progression to the end of a cycle.
2. Requires an EMERGENCY STOP actuator or a circuit to be reset.	2. No reset required.	2. Requires a ZONE STOP actuator or a circuit to be reset.
3. Actuator is hardwired	3. May be hardwired or software controlled.	3. May be hardwired or software controlled.
4. Overrides Zone Stop and Normal Stop.	4. Overrides Zone Stop.	
	5. Includes an END-OF-CYCLE STOP.	

Variation   **Rationale**

7-9 Start Circuits.

Add to Section 7-9(a):
Sensors shall be used to (1) determine the position of equipment elements and parts in process prior to start and (2) control the proper sequencing of the equipment. Self-checking circuits shall be incorporated in the equipment design wherever control malfunctions or improper sequencing may create a hazard to personnel. This circuitry shall give protection against:

Needed in order to specify the manner in which the basic requirements are to be met and provides a means of detection for equipment malfunction.

(1) Failure of devices to function properly.

(2) Improper sequencing in manual and automatic operation.

Add to Section 7-9(b):
Clear-condition interlocks between two pieces of equipment shall consist of normally-open limit switch contacts (to indicate position) in series with a normally-closed relay contact to indicate anticipation of not leaving that position. This circuit shall be permitted to be bypassed with a contact indicating other clear conditions. Where interlocking between separately controlled equipment is required, bypass circuits for the independent operation of each machine shall be provided, except that those interlocks which provide personnel safety or prevent damage to equipment (e.g., position of transfer bar, EMERGENCY STOP, etc.) shall not be bypassed.

Needed in order to specify in detail the manner in which the basic requirements are to be met.

Add Section 7-9(c) thru 7-9(h):
(c) Multiple Start Stations shall be permitted if the following conditions are met;
(1) Interlocking to ensure only one control station active at any one time for starting all motors concurrently.
(2) Pushbuttons shall be provided for individual functions
(3) All start buttons shall be concurrently depressed to initiate the cycle and released between successive operations
(4) Pressure switches alone shall not be used to start or determine sequence of operation

Provides a necessary safety feature for complex machinery.

(d) Control circuits shall be so arranged that during any cycle start no motion will occur unless (1)Cycle selector switch is set for AUTOMATIC and CYCLE START pushbutton is depressed, or (2) Cycle selector switch is set for MANUAL and the pushbutton for a particular function is depressed. If the equipment was previously stopped in mid-cycle, no motion shall occur until one of the sequences shown above has been performed. If two-hand operation is required with the control set for AUTOMATIC, it shall also be required when the control is set for MANUAL, JOG or INCH operation.

Provides a necessary safety feature for complex machinery.

(e) A preset circuit is defined as one that starts the cycle when a set of prearranged conditions are satisfied. PRESET begins at the end of the previous cycle. Preset circuits shall be used only with prior approval of the specifying authority. Timers where used in preset circuits, shall be the nonadjustable type and the total elapsed time shall not exceed the total machine cycle time. For manual operation, PRESET shall be initiated on each cycle by a pushbutton. For automatic cycle, the first cycle shall require operation of a pushbutton.

Provides a necessary safety feature for complex machinery.

Variation	Rationale

(f) Control circuits shall be designed so that when the equipment is in its home or end-of-cycle position, movement of any part of the equipment can be initiated by limit switches only where all of the following conditions are met:

Provides a necessary safety feature for complex machinery.

(1) The control is set for automatic operation

(2) The control is of the holding circuit type and is energized.

(3) The condition for the start of automatic operation is shown by indicating lights.

(g) Rotary cam switches shall not be used as sequence controls on equipment unless the position of the equipment component or the location of the material in process is confirmed by separate controls, such as limit switches, which are use in conjunction with the rotary cam switch.

Provides a necessary safety feature for complex machinery.

Exception: This requirement does not apply to punch presses.

(h) When specified by the specifying authority, provisions for an alarm shall be furnished. This alarm is to sound for approximately five seconds prior to the start of hydraulic pumps or other equipment motions. The specifying authority shall specify the size of contactor or relay to be furnished depending upon the number of signals and electrical current requirements. The control power transformer shall have sufficient capacity to supply this circuit. Installation of alarms is the responsibility of the specifying authority.

Provides for an additional safety feature where deemed appropriate.

7-10 Hold-to-Run Circuits.

Retitle Section 7-10 as 7-10(a):

To add 7-10(b) below.

Add Section 7-10(b):
(b) JOG or INCH circuits shall operate only in the MANUAL mode. The prevention of RUN or AUTOMATIC operation during JOG or INCH shall be accomplished by a selector switch and separate pushbuttons. MANUAL REVERSE shall be considered a jog operation.

To add a necessary safety feature.

7-11 Operating Modes.

Retitle and replace Section 7-11 with 7-11(a):
(a) Each machine shall be permitted to have one or more operating modes (e.g., a teach mode and a run mode for robots) that are determined by the type of machine and its application.

To add 7-11(b) and (d).

Add Section 7-11(b):
(b) Multistation equipment shall be provided with a control station at each station which has a Mode Selector Switch and MANUAL controls for the purpose of tool change, set-up, and maintenance.

Converts the general terms in 7-11 to specifics.

Variation	Rationale

Add Section 7-11(c):

(c) The AUTOMATIC and MANUAL mode shall be initiated only from the main operator's control station. The circuit shall be designed to:

Converts the general terms in 7-11 to specifics.

(1) Deactivate both AUTOMATIC and MANUAL modes on power interruption or on undervoltage condition.

(2) Prohibit defaulting to the other mode when either AUTOMATIC or MANUAL mode is deactivated.

(3) Indicate the activation of AUTOMATIC or MANUAL modes.

Each station of a multistation machine shall be designed so that any one station can be idled or bypassed without interference to control of the equipment or wiring of other stations. Each station shall be equipped with a station NORMAL/BYPASS means. The BYPASS means shall be located at the main operator's console unless otherwise specified by the specifying authority. The NORMAL/BYPASS means shall bypass an individual station when it is placed in the BYPASS mode. The station shall remain interlocked with the transfer automation of any other station. Interlocks shall prevent cycling of the equipment if the station is not in a safe condition. An indication of a station in the BYPASS mode shall be provided.

Provides necessary rules for standardization of a multi-station machine.

7-13 Machinery Door Interlocking.

Retitle Section 7-13 as 7-13(a):

(a) Hinged or sliding doors providing ready access to compartments containing belts, gears, or other moving parts that may expose hazardous conditions shall be interlocked through limit switches or other means to prevent operation of the equipment when the doors are not closed.

To add 7-13(b) and (c).

Add Section 7-13(b):

(b) Equipment requiring safety gates which are not required for equipment cycling shall be interlocked with the AUTOMATIC mode circuitry. Actuation of the safety gate interlock shall perform an appropriate STOP function and deactivate the AUTOMATIC mode circuitry.

Provides a necessary safety feature for complex machinery.

Add Section 7-13(c):

(c) Doors or gates which are powered closed and/or open shall use hold-to-run circuits which require the operator(s) to maintain two buttons throughout the complete cycle or to a point where no hazard to the operator(s) exists.

Provides a necessary safety feature for complex machinery.

7-14 Motor Contactors and Starters.

Retitle Section 7-14 as 7-14(a):

(a) Motor contactors and starters that initiate opposing motion shall be both mechanically and electrically interlocked to prevent simultaneous operation.

To add 7-14(b) and (c).

Add Section 7-14 (b):

(b) The auxiliary contact on any starter or contactor shall not be used in excess of its rating for carrying control circuit loads. Where more capacity is needed, an additional relay or contactor shall be used for this purpose.

Establishes specific limits for auxiliary contacts and specifies an alternate means for larger loads.

Variation	Rationale

Add Section 7-14(c):

(c) In addition to the normally-closed contact on the overload relay, each overload relay shall have an isolated normally-open contact for monitoring purposes. This contact shall close whenever overload relay trips. Where programmable devices are used to control motors, this normally-open contact shall be wired as an input, and used in the internal logic that controls the motor.

Specifies the manner in which overload monitoring is to be accomplished.

7-15 Relays and Solenoids.

Retitle Section 7-15 as 7-15(a):

(a) Relays and solenoids that are mechanically interlocked shall be electrically interlocked.

To add 7-15(b) and (c).

Add Section 7-15(b):

(b) Solenoids shall be individually protected by a fuse or circuit breaker, operated by a single control device, and not connected in parallel with other control devices.

Provides essential details for solenoids not covered elsewhere.

Exception: Indicating devices shall be permitted to be used in parallel with solenoids.

Add Section 7-15(c):

(c) Where solenoids are relay controlled, separate normally-open contacts of the same relay shall be connected in each lead of the solenoid. Not more than two contacts shall control any solenoid. (see sample Elementary Diagram, Appendix D3)

Experience has shown the need to disconnect both leads of a solenoid in ungrounded circuit to prevent unintentional operation due to ground faults.

Exception No. 1: Where a grounded control circuit is specified, a single normally-open relay contact shall be connected in the ungrounded lead of the solenoid.

Exception No. 2: In a grounded control circuit, w here solid-state control is furnished, a single solid-state contact (output) shall be permitted in the ungrounded lead of the solenoid.

Variation	Rationale

7-17 Two-Hand Control Circuits.

Retitle Section 7-17 with:
7-17 Machine Control Circuits.

Replace Section 7-17(a) with:
7-17(a) Machine Guarding Control Circuits.

Provides details necessary for complex machines.

(1) Where used to guard hazardous motion, machine guard control circuits shall:

a. be designed, constructed, and installed such that a single failure within the safeguarding device and its associated circuitry shall not prevent the normal stopping action from taking place, and shall prevent a successive machine cycle.

NOTE: Associated circuitry can include, but is not limited to, electromechanical relays, solid state devices, PES, wiring and input/output devices. Electrically operated presence sensing devices can be used for purposes other than safeguarding of personnel.

b. be located or installed such that the personnel cannot reach into the hazardous area before deactivation of the safeguarding control device and before cessation of hazardous motion.

Not intended for manually operated and visually apparent devices such as "pull plugs".

c. require the concurrent release and concurrent activation of all operator actuated controls used for machine safeguarding.

(2) Two hand control, when used as part of the machine safeguarding control circuit, shall:

a. be located and arranged so that operation by means other than both hands of the operator is prevented.

b. be required for all manual, inch, or jog operations, where required for automatic mode operations.

(3) Presence sensing safeguarding devices (PSSD), when used as part of the machine safeguarding control circuit, shall:

a. have a minimum object sensitivity required for proper safety distances to be achieved as defined in ANSI B11.19.

b. prevent hazardous motion of the machine whenever an obstruction is sensed.

c. incorporate indicators that are visible from the area protected by the presence sensing device.

d. when equipped with a muting system, comply with the control features of this section, and may be such that the muting system can be invoked only when the machine is in its non-hazardous portion of its cycle.

Variation	Rationale

Replace Section 7-17(b) with:

7-17(b) Safeguarding Control Circuits.
(1) Where used to initiate hazardous motion, machine initiate control circuits shall comply with all safeguarding requirements and in addition shall:

  a. be protected against unintentional operation.

  b. not use presence sensing safeguarding devices alone, as an operator actuated control device to initiate hazardous motion.

  c. incorporate an anti-repeat feature for machines that would present a hazard if an unintended repeat cycle should occur.

(2) Two-Hand Control Circuits. Where used to initiate potentially hazardous motion, two-hand control circuits shall:

  a. be protected against unintentional operation.

  b. have the palmbutton contacts connected in series and shall be arranged by design and construction or separation, or both, to require the concurrent use of both hands to initiate the machine operation.

  c. incorporate an antirepeat feature for machines that would present a hazard if an unintended repeat cycle occurred.

    NOTE: See ANSI B11 series standards

(3) Presence Sensing Safeguarding Devices, when used in Presence Sensing Device Initiation (PSDI) applications as part of the machine initiate control circuit shall:

  a. have a minimum object sensitivity of 31.75 mm (1.25 in) or smaller, including blanked areas.

  b. utilize a single sensing field plane for protection.

Devices as described in 7-17(b) 1 thru 3 are not intended to be "safeguarding devices" but could be utilized as such when incorporated with the proper additional control circuit elements.

Delete Section 7-17(c)

Add Section 7-18:
7-18 Programmable Device (PLC) Circuits.

Provides essential details for PLC's which are often the control means for complex machines.

Add Section 7-18(a):
(a)   Only one input device shall be wired to each discrete input module. Indicating devices (e.g., pilot lights) may be wired in parallel with the input point.

Exception: Wiring of multiple input devices to a single input point shall be permitted with the approval of the specifying authority.

Add Section 7-18(b):
(b) Outputs shall control only one device.

Exception: Indicating devices shall be permitted to be used in parallel with the device being controlled.

Add Section 7-18(c):
(c) Input devices shall be wired in the normally-open state.

Exception: Where safety requirements must be met, e.g., stop, reset, flame guards, etc., contacts shall be wired in the normally-closed state.

Add Section 7-18(d):
(d) When applicable, logic and circuitry for lubrication control and cycle monitoring lights and machine cycle time indication shall be incorporated into the Programmable Device. The lubrication logic should precede the normal system control logic.

Add Section 7-19:
7-19 Master Control Relays.

Master control relays are often a part of a control panel and their performance requirements should be included in Chapter 7.

Add Section 7-19(a):
(a) Control circuits having a total of more than eight control devices such as relays, timing relays, and mechanically-held relays shall be equipped with a non-retentive (electrically-held) master control relay. The contacts of this relay shall be inserted in each ungrounded leg of the control circuit in such a manner as to de-energize the entire control circuit.

Add Section 7-19(b):
(b) Control circuits with a non-retentive master control relay shall be designed so that the master control relay:

(1) Removes all control circuit voltage on an undervoltage condition.

(2) Removes all control power upon actuation of the CONTROL POWER OFF button (see Section 7-5).

(3) May be used as the EMERGENCY STOP control circuit if all conditions for EMERGENCY STOP are met.

Variation	Rationale

Add Section 7-19(c):

(c) A master control relay shall be provided with a CONTROL POWER OFF pushbutton or selector switch. This circuitry shall comply with the following requirements:

(1) CONTROL POWER OFF devices shall be hard-wired in electromechanical logic.

(2) CONTROL POWER OFF circuitry shall not use programmable device logic or any other electronic devices.

(3) Except as noted above, all control power shall be removed upon actuation of CONTROL POWER OFF.

Add Section 7-19(d):

(d) Equipment requiring safety blocks shall be interlocked with the master control relay circuit and shall remove all control power when removed from their stored position.

Add Section 7-19(e):

(e) Line side connected devices and auxiliary lighting and receptacles shall not be controlled by a master control relay.

Add Section 7-20:
7-20 Overtime Timers.

Overtime timers are often a part of a control panel and their performance requirements should be included in Chapter 7.

Add Section 7-20(a):

(a) An adjustable cycle overtime feature shall be supplied on single cycle and continuous cycle equipment. This feature shall be designed to include the following:

(1) The cycle overtime timer(s) shall be timing whenever the equipment is "in cycle."

(2) Machines that have long cycles shall be divided into steps, each step being monitored by its own cycle overtime timer.

(3) The cycle overtime timer(s) shall be preset to a value slightly longer (e.g., 10%) than a normal cycle/step.

(4) When a cycle overtime timer times out, AUTO CYCLE shall be deactivated (but not AUTO mode) and a cycle overtime error shall be indicated.

Add Section 7-20(b):

(b) A timer shall be provided to detect excessive wait time on equipment that is "in cycle" and waiting for a non-operator event to initiate action. Excessive wait time shall be annunciated and be a different error than cycle overtime.

Add Section 7-21:
7-21 Motor Related Circuits.

Motor related control circuits and equipment are often a part of a machine control system and their requirements should be addressed in Chapter 7.

Add Section 7-21(a):

(a) Plugging switches or zero speed switches, used to control the application or removal of power, in order that moving parts may be slowed down, stopped, or reversed, shall be provided with features incorporated in the control circuit to (1) prevent the reapplication of power after the completion of the plugging operation; and (2) prevent the application of power through any manual movement of the plugging switch shaft or of the motor or equipment. Timing relays shall not be permitted as the control means for plugging motors.

Add Section 7-21(b):

(b) Shunt and compound wound DC motors shall be equipped with overspeed or field loss protection to prevent excessive motor speed.

Variation	Rationale

Add Section 7-21(c):

(c) Power roll automation drive motors shall be de-energized if no part loading or unloading has occurred within one minute. Drive motors shall be re-energized when the control is set for AUTOMATIC mode, if a part is loaded or unloaded.

Add Section 7-21(d):

(d) Integrally mounted solenoid-operated magnetic brakes shall be permitted to be connected directly across motor leads where the following conditions are met:

(1) The brake coil is rated at full line voltage.

(2) The brake leads are identified as power leads.

Add Section 7-21(e):

(e) When specified by the specifying authority, a nonfusible disconnecting means, meeting the requirements of Chapter 5, shall be provided at each motor that is not within sight of or not readily accessible to the main disconnecting means. The disconnecting means shall include interlock contacts that are connected in the control circuit to de-energize the starter and indicate at the operator's control station when the disconnecting means is open.

Add Section 7-21(f):

(f) Each motor, one horsepower and larger, shall be controlled by one motor starter with individual overload relays of the same manufacturer. Several fractional horsepower polyphase motors may be operated from one starter if each motor has its own overload relay contact wired in series with the starter coil, and the branch circuit is properly protected.

Add Section 7-21(g):

(g) The starting of motors in a group is allowed provided the starting of the motors is sequenced so that any group of motors started simultaneously will not exceed an aggregate of 100 horsepower.

*Limits on **multiple motors** starting with **one starter** or **contactor is limited to size 1**. The **group motor starting** requirements **are allowed for sequenced applications up to** an aggregate of 100 Hp. This is also **coordinated with** section 6-5(b) - Protection.*

Add Section 7-22:

7-22 Lubrication and Hydraulic Systems.

*Lubrication **and hydraulic** systems **are often a part of a** complex **machine and their** relationship to control circuits should be specified.*

Variation	Rationale

Add Section 7-22(a):

(a)   Automatically operated lubrication systems shall be interlocked with the control circuit. Low lubricant level or lubrication system failure shall:

    (1) Prevent the machine or equipment cycle from starting.

    (2) De-energize the machine cycle circuit after the cycle in progress has been completed.

Add Section 7-22(b):

(b) The operational status of the lubrication system shall be indicated at the operator's main control station.

Add Section 7-22(c):

(c) Changing feeds from rapid traverse to feed rate in hydraulic actuated heads shall require a de-energization operation to obtain feed rate.

Add Section 7-22(d):

(d) Each station unit shall have circuits designed so as to return its head unit to its starting position upon completion of its feed cycle independently of other units. Each station unit shall include a HEAD RETURNED indicating light.

Add Section 7-22(e):

(e) For equipment having two or more independent index units, interlocking shall be provided to prevent double loading of any part of the machine even if the equipment is de-energized by the disconnecting means and then is re-energized and restarted.

Add Section 7-23:
7-23 Hydraulic and Pneumatic Valve Control.

Hydraulic and pneumatic valves are often a part of a complex machine and their relationship to control circuits should be specified.

Add Section 7-23(a):

(a) Control of Valves: Electrically controlled (solenoid) hydraulic and pneumatic valves shall be applied in such a manner that in the event of power failure there will be no hazard or damage to the equipment.

(1) The clamping solenoid valve for workpiece clamping applications shall remain energized throughout the working cycle and until the unclamp solenoid is energized.

(2) Clamps actuated by hydraulic or pneumatic means shall be controlled by two-position, double solenoid operated, four-way valves.

Add Section 7-23(b):

(b) Hydraulic indexing mechanisms shall be controlled by three-position, spring centered, double solenoid operated four-way valves.

Add Section 7-23(c):

(c) Pneumatic indexing shall be controlled by two single solenoid, spring offset, three-way valves; or a two-position or three-position four-way valve.

Add Section 7-23(d):

(d) Circuits for two-position hydraulic or pneumatic valves shall be designed to keep their solenoids energized to prevent accidental shifting due to pressure surges or vibration.

Variation	Rationale

Chapter 8: Control Equipment

8-1 Connections.

Add to Section 8-1:

Where used, terminal blocks shall comply with UL 486E. DIN rail mounted terminal blocks shall comply with IEC 947-7-1. Strands of conductors shall be retained by screw-type terminals with captive saddle straps, cage clamps or equivalent means.

Requirements for stranded conductor terminations are added. The intent is to contain the wire strands and to prevent conductor damage. UL Standard covers issues of mechanical integrity, and electrical performance. IEC 947 addresses rail mounted devices.

The connection requirements for other remote devices are covered in Section 14--1(g)

8-3 Manual and Electromechanical Motor Controllers.

Replace Section and Exception 8-3(a) with:

a) Each motor controller shall be identified and shall be capable of starting and stopping the motor(s) it controls and, for alternating current motors, shall be capable of interrupting the stalled rotor current of the motor(s) per the manufacturer's listed ratings. Controllers rated in horsepower shall be used for motors rated 1/8 horsepower or larger.

Where NEMA devices are specified, the requirements of section 8-3 (a)(1) and (c)(1) apply. Where IEC devices are specified, the requirements of 8-3 (a)(2) and (c)(2) apply. (See Equipment data form).

(1) NEMA Rated Motor Controllers
  a) The minimum allowable size is NEMA size 0.
  b) All motor controllers shall be marked and applied in accordance with NEMA ratings. "Dual" rating (i.e. devices marked with both the NEMA ratings and another set of ratings on the same device) are not allowed.
  c) The motor controller shall be sized in accordance with table 8-3 (a)

The added requirements attempt to address the use of IEC rated devices as direct replacement for NEMA designed systems. The requirement simplifies engineering by allowing the IEC devices to be used, but only if they have been tested and rated in accordance with NEMA requirements.

Variation   **Rationale**

Replace Table 8-3(a) with:

Table 8-3 (a) Horsepower ratings for Three-Phase, Single-Speed Full Voltage Magnetic Controllers for non-plugging and non-jogging Duty.

Industry Size of Controller	Continuous Current Rating* Amperes	Horsepower at			Service-Limit Current Rating** Amperes
		200 Volts	230 Volts	460/575 Volts	
00	9	1 1/2	1 1/2	2	11
0	18	3	3	5	21
1	27	7 1/2	7 1/2	10	32
2	45	10	15	25	52
3	90	25	30	50	104
4	135	40	50	100	156
5	270	75	100	200	311
6	540	150	200	400	621
7	810	----	300	600	931
8	1080	----	450	900	1240
9	2250	----	800	1600	2590

Based on ANSI/NEMA ICS-2 1988, Table 2-321-1

* The continuous-current ratings shown in Table 8-3(a) and 8-3(c) represent the maximum rms current, in amperes, the controller may be expected to carry continuously without exceeding the temperature rises permitted by Part ICS 1-109 of NEMA Standards Publication No. ICS 1.

** The service-limit current ratings shown in Tables 8-3 (a) and 8-3 (c) represent the maximum rms current, in amperes, the controller may be expected to carry for protracted periods in normal service.

(2) IEC Rated Motor Controllers

  a) The minimum rating of contactor that may be used is one rated for a continuous full load AC3 current of 18 amperes or greater. (AC-3 Squirrel-cage motors: starting, switching off while running; occasional inching, jogging or plugging but not to exceed 5 operations per minute or 10 operations per ten minutes. All wye-delta and two step auto-transformer starting. or All medium starting duty motors.)

  b) "Dual" rated devices (i.e. ones marked with more than one set of ratings <u>on the same device</u>) are not allowed.

  c) Selection Criteria - The electrical life of IEC motor controllers may vary from that of traditional NEMA starters with equivalent HP ratings. Because of this, the manufacturers published load-life curves should be consulted for selecting IEC motor controllers for applications requiring long life. When electrical life of a contactor is stated by the manufacturer, the data should be accompanied by the following:

  1) The applicability of the data - Is the data typical (meaning representative of more than half) or a guaranteed minimum.

  2) The pass fail criteria used in the test supporting the claimed life.

  3) The mounting and maintenance procedures used to obtain published life.

Replace Section 8-3(c) with:
8-3(c) Jogging - Plugging Duty Motor Controllers.
(1) NEMA Rated Motor Controllers Jogging Plugging operation
Where machine operation requires a motor controller to repeatedly open high motor current, such as in plug-stop, plug-reverse, or jogging (inching) duty, requiring continuous operation with more than five openings per minute or more then 10 in a 10 minute period, the controller shall be de-rated in accordance with Table 8-3(c).

Revised based on NEMA ICS 2-327.22 to add the maximum requirement for a 10 minute period.

(2) IEC Rated Motor Controllers Jogging - Plugging operation
a) The minimum rating of contactor that may be used is one rated for a continuous full load AC3 current of 18 amperes or greater. Contactors for plugging duty shall be sized in accordance with the manufacturers AC-4 application guidelines. (AC-4 Squirrel-cage motors: starting, plugging, inching jogging. or all heavy starting duty motors).
b) "Dual" rated devices (i.e. ones marked with more than one set of ratings <u>on the same device</u>) are not allowed.
c) Selection Criteria - The electrical life of IEC motor controllers may vary from that of traditional NEMA starters with equivalent HP ratings. Because of this, the manufacturers published load-life curves should be consulted for selecting IEC motor controllers for applications requiring long life. When electrical life of a contactor is stated by the manufacturer, the data should be accompanied by the following:
1) The applicability of the data - Is the data typical (meaning representative of more than half) or a guaranteed minimum.
2) The pass fail criteria used in the test supporting the claimed life.
3) The mounting and maintenance procedures used to obtain published life.

Replace Section 8-3(d) with:
(d) Several motors shall be permitted to be controlled by one starter or contactor where all of the following are complied with:

(1) The starter or contactor is not larger than size 1.

(2) Each motor has it's own individual overload relay.

See Chapter 6, Protection, for all requirements concerning motor protection and overloads.

Automotive industry practice has indicated this limited practice is acceptable.

Starting current problems are reduced.

Replace Section 8-4 with:
8-4 Marking on Motor Controllers. Controllers for motors rated 1/8 horsepower or more shall be marked with the voltage, phase, horsepower rating, the rated conditional short-circuit current, and such other data as may be needed to properly indicate the motors and maximum system available faults for which they are suitable with the designated protective device.

Assures markings on motor controllers are acceptable for higher fault current system applications. Markings may be supplementary labels where not appropriate to be marked on the device.

Variation   **Rationale**

8-5 Ratings and Standards.

Add Section 8-5:
8-5 Ratings and Standards. Control circuit devices and three-phase control apparatus shall conform to the applicable ANSI/NEMA or IEC industrial control equipment standards.

Default standards are defined where specific requirements are not identified in component specific documents establishing minimum levels of ratings and performance.

Add Section 8-6:
8-6 Transformers.

(a) Control circuit transformers shall conform to NEMA ST-1-1988, Specialty Transformers, Section 4, Industrial Control Transformers, and shall be Type 1, without integral overcurrent protection. Overcurrent protection, meeting the requirements of Chapter 6, may be mounted on the transformer but overcurrent devices which are internal are not acceptable. The transformer primary winding may be dual rated 240 x 480 volts at 60 hertz with a single secondary winding rated 120 volts. A single primary rated 480 volts is also acceptable.

Standards for control circuit transformers do not exist in the existing NFPA-79 document. The automotive industry standards are presented. Maximum temperature rise, ratings, efficiency, regulation are addressed. Limitations for maximum surface temperatures are included.

(b) Transformers used in combination motor starters shall conform to NEMA ST-1, Section 3, Control Transformers.

(c) Power circuit transformers or drive isolation transformers such as 480/240 or 480/208 volt, single phase or three phase shall conform to NEMA ST-20.

(d) Transformers mounted outside of control enclosures shall have an enclosure rated Type 3R if encapsulated or Type 12 if non-encapsulated. Transformers greater than 7.5 kVA, shall be permitted to be housed in a ventilated enclosure rated Type 2.

(e) Transformers, 10 kVA or less, shall be rated not more than 115° C rise using a 180° C insulation system. Transformers greater than 10 kVA, shall be rated at 150° C rise using a 220° C insulation system. All transformers shall provide a minimum efficiency of 90 percent at full load. The maximum temperature of the hottest exposed surface of an open type, or the enclosure of enclosed transformer, shall be limited to 80° C (176° F) while operating in an ambient of 40° C (140° F). Transformer loading shall be limited to 80 percent of full load to ensure that temperature limitations are met.

(f) All transformers shall provide wire terminations made of materials which are suitable for copper wire connections, for primary and secondary circuit conductors. Copper to aluminum connections shall not be permitted.

Add Section 8-7:
8-7 Control Devices.

Basic automotive industry requirements for these devices included.

(a) General:

(1) Control devices in an enclosure shall be designed to operate continuously and reliably within an enclosure with internal temperatures of 0° C to 55° C for electromechanical devices and 0° C to 50° C for solid state devices.

(2) Control devices external to the control enclosure shall be rated type 12. Where environmental conditions are more severe, devices shall be rated to suit those conditions.

Internal enclosure design temperatures are identified to coordinate device ratings to maximum application temperatures.

(3) The following items shall not be used without written permission from the specifying authority:

1. Stepping switches.

2. Push selector switches for start-stop operation.

3. Latch relays for control of motion.

4. Devices with "make before break" overlapping contacts.

5. Neutral position limit switches.

6. Wire connectors made of aluminum.

7. Maintained position limit switches.

   Exception: maintained position safety cable limit switches.

8. Time delay pushbuttons, selector switches, limit switches, etc.

9. Manifold mounted limit switches.

10. Motor starters with automatic overload reset.

11. Two step sequence limit switches.

(b) Pushbuttons, pilot lights, etc. - See Chapter 11 for requirements.

(c) Relays:
(1) Relays shall have a complete set of contacts (e.g., a four pole block shall have all contacts furnished). At least one spare contact (module) shall be furnished on every relay used in the control circuit.

(d) Contactors for non-motor loads:
(1) Contactors for non-motor loads shall conform to NEMA ICS2-210.

(e) Solenoids:

(1) Solenoids used for actuating valves, brakes and other mechanisms shall have a continuous duty rating.

(2) Electrically energized devices located external to the control enclosure, such as clutches, solenoids and other coil operated devices, shall be enclosed and include an oil tight enclosure for conduit termination and connection purposes. Coils shall have leads which extend a minimum of four **inches outside the enclosure.**

(f) Sensors:

(1) General - Sensors include all devices such as limit switches, proximity switches, pressure switches, temperature switches, etc. which are used to sense or control conditions on equipment.

(2) Limit switches:

1. Limit switches, pressure switches and similar devices shall have separate, isolated normally open and normally closed contacts. (NEMA ICS-2 form Z) Contacts shall be of the quick make/quick break type, nonteasable to an intermediate position.

2. Detailed requirements for limit switches are contained in SAE J1738-1.

(3) Proximity switches:

1. Detailed requirements for proximity switches are contained in SAE J1738-1.

2. Proximity switches (e.g. noncontact inductive, capacitive, magnetic, optical or other types) shall not be used for safety purposes, such as overtravel or for operator guards unless approved by the specifying authority.

Add Section 8-8:
8-8 Presence Sensing Devices. Presence Sensing Safeguarding Devices (PSSD's) when used for safeguarding shall:

(a) be designed, constructed, and installed such that a single failure within the safeguarding device and its associated circuitry shall not prevent the normal stopping action from taking place, and shall prevent a successive machine cycle.
NOTE: Associated circuitry can include, but is not limited to, electromechanical relays, solid state devices, PES, wiring and input/output devices. Electrically operated presence sensing devices can be used for purposes other than safeguarding of personnel.

(b) have the following information stated on the device:
1) minimum object sensitivity
2) maximum response time
3) maximum angle of divergence/acceptance.

(c) provide visual indication or field terminal access to the following signals:
1) power is on/device OK
2) sensing field has been obstructed
3) blanking is active (if offered).

(d) have an enclosure rating suitable for the environment where installed.

(e) meet the requirements of NEMA for electromagnetic interference.

(f) not be adversely affected by external environmental factors such as strobe lights, weld flash, incandescent lights, fluorescent lights, photocells, or sunlight.

(g) not utilize variable analog adjustments for use in different applications. Discreet settings shall be acceptable.

(h) provide an interface that can be monitored for single failure outside the presence sensing device.

Included to establish minimum installation / application requirements for presence sensing safeguarding devices (PSSDs) and ensure that proper safety distances are utilized.

Variable adjustments may allow sensing field changes that may not be readily apparent. Dip switches, detent thumbwheel switches and other like devices are acceptable.

Variation	Rationale

Chapter 9:   Control Enclosures and Compartments

9-1 Type.

Add to Section 9-1(a) (between 9-1(a) and exception):
Enclosures and compartments shall be Type 12 steel, unless otherwise specified in the Equipment Data Form, in accordance with either of the following:

1. NEMA ICS 6, "Enclosures for Industrial Control and Systems"

2. UL 508, "Industrial Control Equipment".

In addition, any corrosion protection process required for the enclosures shall not include the use of cadmium.

Where cooling devices are used, they shall not violate the protection integrity of the enclosure.

CFC and HCFC refrigerants are not permitted.

This paragraph adds the Automotive Industry preference for NEMA 12 enclosures for the majority of applications. A higher or lesser degree of protection may be necessary depending upon the installation and conditions.

Environmental concerns limit the use of cadmium. This note highlights this requirement.

Equipment requiring ventilation and generally not requiring the protection of a Type 12 enclosure (such as batteries, drives compartments, etc.) shall be installed per exceptions (1) or (2).

Note: Prohibition of CFC and HCFC is in anticipation of future limits on use of refrigerants and ensures compliance with the latest environmental requirements.

Add to Section 9-1(b):
Air filters shall be readily available commercial types and sizes. Ventilation openings shall be a minimum of 457 mm (18 in.) above the operating floor line.

Ventilating a NEMA 12 enclosure reduces its rating to NEMA 1.

Variation   Rationale

9-2 Nonmetallic Enclosures.

Retitle and Replace Section 9-2 with:
9-2 Control Enclosure Construction.
Control enclosures shall be made of steel, aluminum, iron, stainless steel, or other   *The revision expands the*
materials as required by the specifying authority. [See Appendix G, 5(b).]   *requirements for enclosure*
                                                                                        construction based on typical
                                                                                        automotive installations.

Where corrosion protection beyond normal requirements is needed, non-metallic   *Performance requirements*
enclosures identified for the purposes shall be permitted if they meet the   *for non-metallic enclosures*
requirements of UL 508. For grounding provisions, see Section 17-3.   *are referenced to an accepted*
                                                                                        industry standard.

Add Section 9-2(a):
   (a) Control enclosures shall provide a means to divert all liquids from entering   *Depending on gasket design,*
the enclosure or having continuous contact with the door gasket.   *liquid contact with gasket*
                                                                                        material causes failures and
          Exception: motor control centers and factory built single motor controllers *leaks with knife-edge*
                                                                                        enclosure flanges. Provides
                                                                                        extra degree of protection
                                                                                        against liquids entering the
                                                                                        enclosure.

Add Section 9-2(b):
   (b) Holes, cutouts and provisions for mounting through the enclosure walls shall
be closed or sealed to maintain the original protection category of the enclosure.

Add Section 9-2(c):
   (c) Subplates shall be self aligning, easily removable, designed to minimize   *12 gauge minimum sub-plate*
flexing and shall install firmly and securely to the enclosure. Subplate thickness   *requirement provides 2*
shall be a minimum of 2.7 mm (.106 in.) for mounting of components by drilling   *threads of engagement in*
and tapping the subplate.   Maximum screw size is 1/4-20 UNC unless other   *drilled and tapped holes.*
provisions are made to provide proper thread engagement.

Add Section 9-2(d):
   (d) Supports are required for subplates having a surface area of more than 15,484   *Subplates supports are*
sq. cm (2400 sq. in.). Supports shall support the weight of the subplate and the   *required for large enclosures*
mounted components. Supports shall also support the subplate during installation   *for ease of installation of the*
into the enclosure.   *subplate, and to help support*
                                                                                        the weight of the components
                                                                                        and sub-plate combined.

Add Section 9-2(e):
   (e) Enclosures and subplates shall be free of burrs and sharp edges.

Variation	Rationale

9-4 Wall Thickness.

Add to Section 9-4:
The metal thickness shall be in conformance with the requirements of UL 508, but with a minimum of 1.52 mm (.06 in.).

This better defines the requirements and accepts recent improvements in control enclosure design and technology (Minimum thickness of 16 gauge based on old standard thicknesses)

The thickness of sheet steel used for walls and doors of enclosures or compartments shall be as shown in Table 9-4.

Exception: If a supporting frame or equivalent reinforcement is used, the minimum enclosure wall thickness for areas over 7742 sq. cm. (1200 square inches) shall be 1.5 mm (0.053 inch) if made of sheet steel.

Larger enclosures need greater wall thickness if not supported by a frame. The enclosures dimensions in UL 508 provide for a thinner wall construction then the automotive environments dictate.

Add Table 9-4:
Table 9-4 Metal Thickness for Walls and Doors of Enclosures or Compartments

Maximum Area of Any Surface, Sq. In.	Maximum Dimension, Inches	Minimum Thickness (Nominal) of Metal, Inches
Less than 360	18 (457 mm)	0.060 (2.0 mm)
Greater than 360 and less than 1200 (7742 sq. cm)	48 (1220 mm)	0.075 (2.0 mm)
Over 1200	60 (1525 mm)	0.106 (2.7 mm)

9-5 Dimensions.

Add to Section 9-5:
The height and width of the door opening shall be greater than the corresponding height and width of the subplate.

Required to allow the sub-plate to be installed and removed easily.

9-6 Doors.

Add Section 9-6(a) after the exception:
 (a) Door fasteners on enclosures and compartments with door openings less than 1016 mm (40 in.) tall shall be designed to seal the door tightly around its perimeter with either captive fasteners or vault type hardware which latch at the top and bottom.

Two latch points for small doors and three for large doors required to ensure NEMA 12 and 13 protection.

Variation	Rationale

Add Section 9-6(b):

(b) Door fasteners on enclosures and compartments with door openings 1016 mm (40 in.) tall or more shall be designed to seal the door tightly around its perimeter with either captive fasteners or vault type hardware which latch at the top, center and bottom.

Two latch points for small doors and three for large doors required to ensure NEMA 12 and 13 protection.

Add Section 9-6(c):

(c) Vault type hardware shall be used on the door enclosing the disconnecting means and shall be interlocked with the disconnecting means (see Section 5-9).

On the disconnect enclosures all three latch points should be interlocked with the disconnect handle.

Add Section 9-6(d):

(d) Doors on enclosures shall be designed to UL 508 requirements for door rigidity. Doors shall have welded, brazed, or formed corners.

Door flexing should be minimized to maintain Type 12 protection. Compliance to UL 508 meets the requirement for reinforcement.

Add Section 9-6(e):

(e) Door swing shall be a minimum of 120 degrees and shall be such that ready access to the disconnect handle is not blocked. Doors shall be easily removable.

120 degree door swing is necessary to facilitate unobstructed servicing and the use of swing frames for electronic components. The disconnect handle should be accessible with the door open. Doors should be capable of being removed for assembly and maintenance operations.

9-7 Gaskets.

Add to Section 9-7:

Door sealing gaskets shall be at least 6.4 mm (1/4 in.) thick and be held firmly in place with oil resistant adhesive or equivalent. Gaskets shall meet the requirements of UL 508 Type 12.

Requirement for oil resistant adhesive is added to insure compatibility. Equivalent gasket types like foamed-in place gaskets may be acceptable. The thickness requirement is to insure a minimum level of quality for enclosure gaskets. Gasket properties are defined by referencing UL 508.

Variation	Rationale

9-9 Interior Finish.

Retitle and Replace Section 9-9 with:
9-9 Finish.
The exterior of the enclosure shall include a protective finish. The interior of the enclosure, including doors, shall be finished in light color (see definition Appendix A). The subplate shall be finished in gloss white or gloss orange. [See Appendix G, 5(d).)

Retitled to include both interior and exterior finish. Interior reflectance requirements are to improve visibility and allow safe working conditions when servicing.

The intent is not to be too restrictive in the definition of internal enclosure colors.

9-10 Warning Mark.

Retitle and Replace Section 9-10 with:
9-10 Warning Signs on Enclosures.
All enclosures which do not clearly show that they contain electrical devices shall be marked with a black lightning flash on yellow background within a black triangle, shaped in accordance with the graphical symbol 417-IEC-5036, the whole in accordance with ISO 3864 symbol No. 13: Safety colors and safety signs.

Changed to reference industry standard symbol.

9-11 Print Pocket.

Add to Section 9-11:
When door size allows enough space, the print pocket shall be at least 305 mm (12 in.) wide x 305 mm (12 in.) high and of 25 mm (1 in.) minimum depth to accommodate all electrical diagrams.

Variation	Rationale

Chapter 10   Location and Mounting of Control Equipment

10-1 General Requirements.

Add Section 10-1(c):

(c) Components shall be mounted to provide mechanical clearances sufficient for mounting, wiring, adjustment, testing and replacement. Each component shall be mounted to provide heat dissipation consistent with the temperature rating of the component, adjacent components and conductors. Each component shall be arranged and oriented so that the device circuit identification may be determined without moving the component or its wiring.

The existing text addresses adjustments and maintenance. SAE additionally addresses heat dissipation, identification, mounting, wiring, testing and replacement.

Add Section 10-1(d):

(d) The following may be mounted on the top of the enclosure and shall allow for and not interfere with installation of the main power feed.

  (1) Power factor correction capacitors.

  (2) Control transformers larger than 2 kVA.

  (3) Line conditioning and regulating transformers of 500 VA or larger.

These devices shall be mounted so that no portion of the equipment immediately above the door opening and less than 2.13 meters (84 inches) from the floor projects more than 15.2 cm (six (6) inches) in front of the door frame.

To clarify which components may generally be mounted on top of the main control enclosure to eliminate the heat generated from inside the enclosure also provides guidance to mounting location to provide adequate headroom clearance.

Add Section 10-1(e):

(e) Devices such as pushbuttons, selector switches, meters, operator adjustable controls and pilot lights may be door-mounted; however, they shall be wired from terminal strips on the subplate and shall not exceed 120 volts. Wiring between devices on the door is acceptable.

Clarify acceptable practices for mounting devices on enclosure doors.

Add Section 10-1(f):

(f) Electronic devices and the associated wiring shall be segregated from the electromagnetic control and power wiring to minimize electromagnetic interference.

Add Section 10-1(g):

(g) Plug in devices shall be mechanically secured in place with captive fasteners and keyed for proper insertion.

Variation   Rationale

Add Section 10-1(h):

(h) Working Clearances: Sufficient space and access shall be provided about all electrical equipment and enclosures to permit safe operation and maintenance of such equipment. In all cases, the work space shall permit at least a 90-degree opening of the door or hinged panel.

NOTE: Reference NFPA 70 (NEC) 110-16 for requirements for working space in the direction of access to live parts operating at 600 volts nominal, or less to ground and likely to require examination, adjustment, servicing, or maintenance while energized. Reference NFPA 70 (NEC) 670-5 **Clearance** for additional requirements.

The basic work practice in the Automotive Industry is that the equipment is to be "deenergized" to do servicing or maintenance. The working clearances required by the NEC Section 110-16 do not apply under these conditions, however, ergonomic considerations shall be applied to provide appropriate access and space for the tasks involved.

The working clearances that exist determine the tasks that may be permitted to be performed (while "live parts" are present) and the appropriate personal protective equipment and materials required. Diagnostic and troubleshooting testing with appropriate test instrumentation may be permitted while the equipment is "energized".

10-2 Control Panels.

Replace Section 10-2(a) above Exception:

All devices connected to supply voltage shall be grouped above or to the side of devices connected only to control voltages.

The term "Control Panel": as used in the Automotive Industry is interchangeable with the term "Control Enclosure". The term "Control Panel" or "panel" as used in the existing NFPA 79 - 1991 text is commonly referred to as the "subplate" in existing Automotive Industry Standards. The SAE definitions in Appendix "A" are modified to reflect this use. This section deals with equipment inside the "control enclosure" on the "subplate".

Variation	Rationale

Add to Section 10-2(b) above Exception:
Separately grouped terminals shall also be used for remote interlock wiring and signal circuits.

Added to ensure isolation and for ease of identification and troubleshooting.

Add to Section 10-2(c):
The blocks shall not be mounted above each other in a plane perpendicular to the panel. Terminals shall not be mounted in raceway or wireway. Ten percent spare terminals shall be provided on each subplate of every electrical enclosure and compartment. A minimum of eight spare control terminals and three spare power terminals shall be provided.

Spare terminals are generally required for future additions and repairs.

Add to Section 10-2(f):
Subplate-mounted rails or mounting tracks are acceptable. Relays shall be affixed to a mounting track.

Add Section 10-2(h):
(h) Spare space should be a minimum of 20 percent of the total area on the subplate. Like components on the subplate shall be grouped and spare space left for expansion near each group. All components, including spares, shall be identified in accordance with the requirements of Section 2-7.

Exception: This does not apply to combination starters.

Spare space on the subplate is required for expansion or changes. Generally the spare space should permit at least two units of each type for expansion. Additional spare space should be appropriately distributed to bring the total to 20 percent.

Add Section 10-2(i):
(i) Components, normally subplate mounted for any one machine, shall be mounted in one enclosure or compartment.

Logical mounting and working of components facilitates troubleshooting when problems occur.

Add Section 10-2(j):
(j) Subplate mounted control components, such as relays, starters and contactors shall be mounted in numerical order from left to right and top to bottom.

Variation

Rationale

Add Section 10-2(k):

(k) Component mounting:

(1) When components are mounted with studs or machine screws, the mounting plate shall have a minimum thickness in accordance with Table 10-1.

(2) For thicknesses or materials not listed in Table 10-1, the thread engagement shall be sufficient to provide clamping forces equivalent to thread engagement requirements specified in Table 10-1.

(3) Sheet metal drive screws shall not be used to mount components. (Thread cutting screws may be used where the minimum thread engagement is in agreement with Table 10-1.)

(4) Rivets, welds, solders or bonding materials shall not be used to mount components.

Requirements for component installation needed to ensure secure mounting of components while allowing for replacement when required.

Add Section 10-2(l):

(l) Swing out panels located between the enclosure or compartment door and the subplate shall not be used.

Exception: Electronic panels may be of the swing out or sliding type for servicing (Ref. 10-3 (b)).

The use of swing out panels may increase the potential for component failure during normal operation as well as troubleshooting.

Variation   Rationale

Add Table 10-1:
Table 10-1 Machine Screw Usage

Nominal Machine Screw or Stud Size	Minimum Thread Engagement In Aluminum		Minimum Thread Engagement In Steel	
	Inches	Millimeter	Inches	Millimeter
No. 4*	0.075	1.90	0.050	1.27
No. 6	0.094	2.39	0.062	1.59
No. 8	0.094	2.39	0.062	1.59
No. 10	0.125	3.18	0.083	2.11
No. ¼"	0.150	3.81	0.100	2.54

*Number four screws may be used only when components require this size. Sizes smaller than number four shall not be used.

10-3 Subpanels and Electronic Subassemblies.

Retitle Section 10-3 as Section 10-3(a) and add:
Refer to Rationale for Section 10-2(a).

Add Section 10-3(b):
(b) Swing frames and/or swing out panels may be used, provided swing is more than 110 degrees and the voltage does not exceed 150 volts. Wiring shall not inhibit swing. Components mounted behind swing frame shall be accessible when swing frame is open. Only wireway shall be mounted directly behind hinged side of swing frame.

Swing frames allowed to permit more efficient use of control panel space. Voltage limitation on swing frame panels is added to reduce hazards while working inside the enclosure.

Add Section 10-3(c):
(c) Printed circuit boards shall be mounted in the vertical plane.

Required to ensure adequate cooling.

Add Section 10-3(d):
(d) Components shall be mounted for ease of replacement and maintenance after assembly. Controls and adjustments for maintenance personnel shall be separately located from those required by operating personnel.

Required to address the issue of operating personnel having appropriate access to adjustments.

10-5 Clearance in Enclosures and Compartments.

Add to Section 10-5(a):
Refer to Rational for Section 10-2(a).

Variation	Rationale
Add to Section 10-5(b): The clearance requirements also apply to covers or doors for the enclosure. The minimum clearance for voltages between 250 volts and 600 volts shall be increased to 25 mm (1 inch).	*Defines minimum clearances to enclosure doors or covers and provides additional clearance for voltages 250 volts and above. There is no intent to require modifications of the manufacturer's mounting details if the clearance to the "subplate" is less than specified.*
Add Section 10-5(c): (c) Where wire channels are used, the distance between components and the wire channel shall be a minimum of 25 mm (1 inch).	*The 1 in. requirement permits access for component wiring, maintenance, and identification of wire numbers.*
Add Section 10-5(d): (d) Components shall not be mounted directly above or behind the disconnecting means.	*Supplements the requirements of Chapter 5-8(a).*
Add Section 10-5(e): (e) The bottom of the lowest subplate mounted device shall not be less than 457 mm (18 inches) above the servicing level. In no case shall the top of subplate mounted components be more than 2134 mm (84 inches) above the servicing level.	*Provides convenient working access by defining the minimum and maximum heights.*

10-6 Machine Mounted Control Equipment.

Add to the Title of Section 10-6: (General requirement for field mounted devices)	*The title change is to clarify that this applies to field mounted devices.*
Add Section 10-6(d): (d) The manufacturer's standard mounting details shall not be modified or altered.	*For replacement purposes, device mounting should not be altered.*
Add Section 10-6(e): (e) Limit switch actuators shall be designed and applied in accordance with the switch manufacturer's specifications for travel, fly back and other related characteristics. Reverse flow of parts, pallets, etc., shall not damage the actuator or switch, and shall not require any repositioning of the actuator or switch.	*Provides additional guidance for the application of limit switches.*

Variation   Rationale

Chapter 11: Operator's Control Stations and Equipment

11-1 Pushbuttons, Selector Switches, Indicating Lights.

Replace Section 11-1(a) with:

(a) Interface devices, including pushbuttons, selector switches, indicating lights, illuminated pushbuttons, thumbwheel switches, membrane type switches or keypads, and other operator interface devices shall be identified oiltight Type 13. These requirements shall also apply to operator interface devices such as LED, LCD, CRT, gas discharge and neon indicating devices.

Additions were made to include more interface devices and to ensure the integrity of the enclosure.

Replace Section 11-1(b) with:

(b) Pushbutton operators, indicating light lenses and illuminated pushbutton lenses shall be color coded in accordance with Tables 11-1(a) and 11-1(b).

Tables 11-1(a) and 11-1(b) provide coordination with the IEC basic color code standards.

Replace Table 11-1 with Table 11-1(a):
Table 11-1(a) Color Coding for Pushbuttons

Color	Meaning	Explanation	Examples
RED	Emergency	Actuate in case of hazardous condition or emergency, Initiation of a stop function.	Emergency Stop, Master Stop, Off. Stop, Stop of one or more motors.
YELLOW (AMBER)	Abnormal	Actuate in case of abnormal condition, Intervention to suppress abnormal condition or restart interrupted automatic cycle. Return machine elements to a safe position.	Emergency Return, Return slide(s), Horn reset.
GREEN	Safe	Actuate in case of safe situation or to prepare normal conditions. General start, start of cycle.	Master Start, Control On, Start of one or more motors, Cycle Start, Advance Slide(s).
BLACK	Safe	Actuate in case of safe situation or to prepare normal conditions. Use in place of green as needed.	Start of one or more motors, manual functions, inch, jog, lamp test, test cycle.
BLUE	Mandatory	Actuate in case of condition requiring mandatory action.	Reset function
WHITE or GRAY	No specific meaning		

Variation   Rationale

Add Table 11-1(b):
Table 11-1(b): Color Coding for Indicating Lights

Color	Meaning	Explanation	Examples
RED	Emergency, danger or alarm	Hazard, fault condition, protective device has stopped the machine. Immediate attention required.	Faults, pressure/temperature out of safe limits, overload, E-stop device tripped, lube fault, overtravel, battery failure.
YELLOW (AMBER)	Abnormal condition; Impending critical condition	Attention, caution, marginal condition requiring monitoring or intervention. Mechanical devices in forward position.	Pressure/temperature etc. exceeding normal limit; slide advanced, elevator raised, manual mode, battery low, over cycle.
GREEN	Normal, safe condition, ready	Authorization to proceed, indication of normal working limits. Mechanical devices in returned position.	Machine ready, auto mode, motors on, slide returned, elevator lowered, cycle complete, all stations in auto, control power on, Hydraulic pressure OK, parts present.
BLUE	Optional	No specific meaning, may be used to prompt operator.	Enter preset value, machine in bypass.
WHITE	Neutral	Any meaning may be used where required whenever doubt exists, or duplication exists with the use of RED, YELLOW, GREEN, or BLUE.	Ground detector lights, full depth, lube cycle on, machine-in-cycle.

Note: Illuminated pushbutton actuators shall be color coded in accordance with Table 11-1(b) if the illuminated pushbutton is replacing an indicating device. If the illuminated pushbutton is being used in an application where the light is used to prompt the operator, follow Table 11-1(a). In case of difficulties in assigning an appropriate color, WHITE shall be used. When illuminated emergency stop actuator(s) are used, the color RED shall not depend on the illumination of its light.

Add to Section 11-1(d):
These stop function operators shall be unguarded.   *Added the automotive requirement for unguarded operators for stopbuttons.*

Add to Section 11-1(e):
(e) These operators shall be of the guarded type. A guarded pushbutton is a   *Guarded type buttons are required for start functions.*
pushbutton so constructed that when properly mounted, the chance of inadvertent
operation will be minimal. Recognized constructions include: recessed, shrouded,
shielded, covered, and lockable. Refer to NEMA ICS 2 for definitions and
additional information.

Variation	Rationale

11-2 Emergency Stop Controls.

Add Section 11-2(c):

(c) An "Emergency Pull Cord(s)" shall be permitted to replace emergency stop pushbuttons.

Pull cord(s) shall operate manual reset switch(es) and shall be accessible from any position where an emergency stop function is required. The switch shall detect broken or disconnected pull cords and initiate the stop function.

The cord shall be installed between 0.9 and 1.8 meters (3 and 6 feet) from the operating floor and supported at frequent intervals, (not to exceed 3 meters (10 feet)). Supports shall be designed and installed to minimize pull forces and cord wear.

Cords shall be installed so that they do not contain more than the equivalent of two quarter bends, (180 degrees, total), if the cord is anchored on one end and connected to a manual reset switch on the other end. Four quarter bends, (360 degrees, total), is permitted if the cord is connected to manual reset switches on both ends. The cord length shall not exceed 23 meters (75 feet).

Cords shall consist of a minimum of 3 mm (1/8 inch) diameter, 19 strand airplane type cable to ensure proper operation and durability. A corrosion resistant coating is required and shall be red.

Automotive Industry requirements for "Emergency Pull Cord" are added to provide directions for installation when these devices are used. The determination of their use is based on machine or system requirements including ergonomics.

11-3 Foot-Operated Switches.

Add Section 11-3(b):

(b) Contacts shall be snap action.

Add Section 11-3(c):

(c) The use of foot switches shall require approval by the specifying authority.

The use of foot-operated switches should be limited.

11-4 Control Station Enclosures.

Add to Section 11-4:

Where six or more operator interface devices are required, terminal strips shall be furnished in the enclosure, and spare space shall be provided for 25 percent or a minimum of two additional pushbuttons or indicating lights of the same size.

Spare spaces are required for field modifications. "Space" requirement does not mean "a hole" only room for added devices. These spaces may be eliminated for installations where modifications are not expected, with permission of the specifying authority.

11-5 Arrangement of Control Station Components.

Variation   Rationale

Add Section 11-5(a):

(a) Control pushbuttons, indicating lights, selector switches, panel meters, etc. shall be arranged so the automatic controls are grouped and separately spaced from the manual controls. All controls shall be arranged in a logical order in accordance with the sequence of operation. Ground detector indicating lights shall be mounted on the front of the main control enclosure (refer to Chapter 17 for requirements concerning grounded circuits).

Present standard does not address the logical placement of components on the control station. Uniformity of design minimizes operator confusion.

Add Section 11-5(b):

(b) The use of touch type operator interface devices to perform motion or process control functions shall require approval of the purchasing division and shall conform to the circuit design requirements of Chapter 7. Operator Interface Terminals or touch screens shall be arranged similar to the equivalent buttons and lights.

This specific approval is to ensure the consistent application of operator interface stations, not to limit their use.

Add Section 11-5(c):

(c) Devices such as potentiometers and selector switches having a rotating member shall be mounted so as to prevent rotation of the stationary member.

Rotating devices may change position if not secured in place by a detent, key or equivalent. Friction alone is not sufficient to meet this requirement.

Variation	Rationale

Chapter 12   Accessories and Lighting

12-1 Attachment Plugs and Receptacles External to the Control Enclosure.

Add Sections 12-1(d) through 12-1(g):

(d) A locking feature shall be provided for plugs and receptacles to prevent accidental disconnections. Automatic locking upon full insertion is preferred.

Provides clarification to existing requirements and adds additional requirements considered desirable in the automotive industry..

(e) Plugs used for circuits 300 volts and above shall not be removable or insertable when voltage is applied to the receptacle.

Receptacles for power devices (above 300 volts) are required to be interlocked to prevent insertion and removal while energized to reduce the potential risk of fault.

(f) Where more than one plug and receptacle is used in a control system, they shall be identified. They shall also be mechanically coded, polarized, or otherwise arranged to prevent incorrect insertion or interchange.

Exception: The above requirements do not apply to plugs and receptacles to be used for accessory equipment (e.g. hand-held power tools, test equipment).

Disconnected plugs or receptacles shall not have any voltage present on exposed pins.

As required by NFPA 2-8(e) all components should be identified.

(g) Enclosures which house programmable electronic devices shall be provided with a minimum of one (1) duplex receptacle protected at 5 amperes to use with support equipment. This receptacle(s) shall be located adjacent to the programmable electronic devices as well as adjacent to any remote programming ports. Control circuits and devices shall not derive control power from this receptacle(s).

Programmable equipment requiring the use of auxiliary programming devices require the immediate availability of a duplex receptacle during troubleshooting and maintenance of equipment. This requirement insures that these receptacles will be readily available inside the enclosure.

Variation   Rationale

12-3 Control Panel, Instrument, and Machine Work Lights.

Replace Section 12-3(b) 5:

    5. Transformer disconnect devices consist of a transformer, with integral disconnecting overcurrent protection, mounted in an enclosure adjacent to the main disconnecting means. These devices are used to provide separate power to enclosure lighting, work lights, and electronic equipment as required by other applicable sections of this specification.

    Transformer/disconnect devices shall consist of the following:

    a. A NEMA Type 1 enclosure when mounted inside another enclosure, and NEMA Type 12 enclosure when mounted externally.

    b.. A lockable disconnecting means meeting the requirements specified in Chapter 5, except items 5-5-exception, 5-8(c), 5-10(a)(e)(f), and 5-11 do not apply.

    c. Integrated primary and secondary fusing in accordance with the requirements of Chapter 6.

    d. Primary fuses shall be mounted with fuse clip line terminals mounted as close as practical but in no case greater than four inches from the load side terminals of the disconnect.

    e. Integral transformer(s), meeting the requirements of Section 8-7(a).

    There shall be no exposed live parts when the disconnecting means is in the open position.

Additional requirement considered necessary for maintenance and service of equipment in automotive industry. Line side fed transformers without a separate disconnect device are not acceptable. This minimum specification identifies the basic requirement determined to be acceptable for automotive industry applications.

Delete Section 12-3(b) 6.

Replace Section 12-3(f) with:

    (f) Work Lights. When work lights are required [See Appendix G-20(a)], they shall be provided with an ON-OFF switch conveniently located on the equipment. Lamp holders shall not incorporate a switch or receptacle. Work lights used in wet locations shall be provided with ground fault protection.

Additional requirements considered desirable in the automotive industry.

Add Section 12-3(i):

    (i) Control enclosure lighting: When specified on the Equipment Data Form (Appendix G-20(b)), fluorescent interior lighting shall be provided in enclosures of 305 mm (12 inches) or more of depth, accommodating a panel having an area of 9678 sq. cm (1500 sq. in.) or more, and in enclosures accommodating a panel having an area of 23,227 sq. cm (3600 sq. in.) or more, regardless of depth.

Additional requirements considered necessary for maintenance and service of equipment in the automotive industry.

Variation	Rationale

Chapter 13   Conductors
13-1 General.

Add to Section 13-1(a):
5. TFN, TFFN - Moisture-, Heat-, and Oil-Resistant PVC Insulation/Jacket 60° C (140°F) Wet Locations.

TFN and TFFN are allowed for conductor sizes less than 14 AWG.

Replace Section 13-1(b):
Multiconductor flexible cords, Type SO, STO, STOW, SOW, SOO, SOWA, and SOOWA shall be permitted. Other Types of flexible cords which have equivalent or superior characteristics and ratings shall be permitted.

250 volt insulation is not permitted.

Added additional cable types for flexible service.

Add Section 13-1(d) :
(d) Cross-Linked Synthetic Polymer wire, type XHHW (moisture-, and heat-resistant) shall be permitted for sizes 8 AWG. and larger.

The reduced overall diameter of type XHHW wire permits the use of smaller conduit sizes for power wiring.

13-2 Conductors.

Add to Section 13-2(a):
Table 13-2(a) shall apply to single conductors and the individual conductors of multiconductor cables, shielded and unshielded.

Add to Section 13-2(b):
Table 13-2(a) shall apply as shown except the footnote (\) shall be replaced with: Nonflexing construction shall not be permitted for flexing service. All flexing applications shall utilize conductors conforming to ASTM Designation B-174, Class k (1980) or ASTM Designation B-172, Class K (1980).

Nonflexing construction shall not be permitted for flexing service.

Replace Section 13-2(c):
Appropriately sized solid conductors shall be permitted on individual devices that are purchased completely wired (e.g. motor starters, etc.). Appropriately sized solid copper bus bars shall be permitted for grounding or device interconnection.

Solid wires may be subject to premature failure due to vibration especially if nicked or damaged during installation.

Replace Section 13-2(e) with:
(e) Conductors in single or multiconductor shielded cables shall be stranded copper not smaller than No. 22 AWG. If a foil shield is used around the conductor, it shall provide a continuous conduction surface in the presence of bending and flexing. A continuous drain wire shall be provided. The shields and drain wire shall be covered with an oil and moisture resistance jacket.

Minimum wire size for electronic assemblies is 22 AWG. and provide adequate conductivity and mechanical strength.

Variation
_____

_____

13-3 Conductor Sizing.

Replace Section 13-3(d) (including (1) and (2)) with:

(d) Internal wiring of electronic, static and precision devices and shielded cable assemblies used with these devices .... No. 22

Minimum wire size for electronic assemblies is 22 AWG.

(Note: Existing exceptions in 13-3 remain.)

Add Section 13-3(e):

(e) All wire sizes noted above are AWG.

Variation	Rationale

13-5 Conductor Ampacity.

Replace Table 13-5(a) and Notes with:

Table 13-5(a) Conductor Ampacity Based on
Copper Conductors with 90°C Insulation
in an Ambient Temperature of 30°C

Conductor Size AWG	Ampacity In Cable, Raceway and Control Enclosure
30	0.5
28	0.8
26	1
24	2
22	3
20	5
18	7
16	10
14	15
12	20
10	30
8	50
6	65
4	85
3	100
2	115
1	130
0	150
2/0	175
3/0	200
4/0	230
250	255
300	285
350	310
400	335
500	380
600	420
700	460
750	475
800	490
900	520
1000	545

The source of the ampacities in this table is the 75°C column of the 310-16 Table of the *NEC*.

NOTE 1: Wire types listed in Section 13-1 shall be permitted to be used at the ampacities as listed in this table.

NFPA 79 Section 3-3 states the equipment shall be capable of operating in an ambient temperature range of 5 to 40°C. Table 13-5(a) is for an ambient of 30°C.

The new table eliminates the NEC 310-17, 60°C column in the control enclosure. This column, because of it's higher ampacities, may not be appropriate for present day starters.

Since nearly all devices are now rated for connection to NEC 310-16, 75°C selected wire, the 60° column was eliminated.

In addition, derating tables were included with the table for convenience, (Table 13-5(b)).

Variation　　　　　　　　　　　　　　　　　　　　　　　**Rationale**

Table 13-5(b)
Ampacity Derating Tables

For ambient temperatures other than 30°C (86°F), multiply the allowable ampacities shown above by the appropriate factor shown below.		Where the number of current-carrying conductors in a raceway or cable exceeds three, the allowable ampacities shall be reduced as shown in the following table:	
Ambient Temp. °C	**Factor**	**Number of Current-Carrying Conductors**	**Percent of Values in Tables as Adjusted for Ambient Temperature if Necessary**
21-25	1.05	4 through 6	80
26-30	1.00	7 through 9	70
31-35	.94	10 through 20	50
36-40	.88	21 through 30	45
41-45	.82	31 through 40	40
46-50	.75	41 and above	35
51-55	.67		
56-60	.58		
61-70	.33		

Chapter 14   Wiring Methods and Practices

14-1 General Requirements.

Replace 14-1(a) "color" requirement with:

BLACK-line and load circuits, ac greater than 120 volts (nominal), or dc above 50 volts.

RED-AC control circuits, 120 volts and lower.

BLUE-DC control circuits, 50 volts and lower.

YELLOW-All control circuits (DC above 50 volts or AC above 30 volts) which may remain energized when the main disconnecting means is in the off position. Each such conductor shall be yellow throughout the entire circuit, including wiring in the control panel and external field wiring.

NOTE: The international and European standards require the use of the color ORANGE for this purpose. (See IEC 204-1 for specific requirements.)

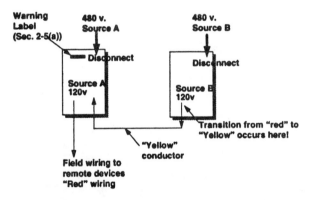

GREEN - (with or without a yellow stripe) equipment grounding conductors (non current carrying).

NOTE: The international and European standards require the use of the bicolor GREEN-AND-YELLOW for this purpose. (See IEC 204-1 for specific requirements.)

WHITE OR NATURAL GRAY -grounded (current carrying) AC circuit conductors.

WHITE with a BLUE STRIPE - grounded (current carrying) DC circuit conductor. NOTE: The international and European standards require the use of the color LIGHT BLUE for this purpose. (See IEC 204-1 for specific requirements.)

WHITE with a YELLOW STRIPE - grounded (current carrying) AC circuit conductor which may remain energized when the main disconnecting means is in the off position.

(The Exceptions 1-6 in Existing NFPA 79 remain unchanged.)

The intent of wire color coding is to identify the voltage potential and separately energized sources, for the safety of servicing personnel.

The point where the transition from the "red", "white", or "blue" conductor to the "yellow" conductor should occur at the voltage supply source.

WHITE WITH BLUE STRIPE was added to harmonize with current standards practices.
WHITE WITH YELLOW STRIPE added to identify remotely sourced neutral conductor.

Variation	Rationale

Delete 14-1 (b) *Exception* **and replace with the following:**
Note: Electrical connections to motors, solenoids and similar devices with integral leads shall be made in accordance with the requirements of Section 14-4.

The connection of utilization devices requires the ability to conveniently disconnect and remove for maintenance or replacement. Permanently spliced connections for these devices are not acceptable.

Add to 14-1 (c)
Terminals on terminal blocks shall be numbered in numerical ascending order, starting from top to bottom, from left to right.

Facilitates maintenance by providing logical organization and arrangement.

Exception: Terminals for remote interlock wiring shall be grouped separately.

Add to 14-1 (e)
Handwritten tags shall not be used. Wire identification tags shall consist of a single tag, and identification text shall be permanent.

Legible, permanent tags are necessary to facilitate maintenance and troubleshooting.

14-2 Panel Wiring.

Add to Section 14-2 (a)
Wiring to door mounted devices may be bundled in lieu of installation in wire duct. Wiring to devices on the enclosure doors shall not interfere with the devices on the door, components on the subplate, or operation of the door.

Additional clarifications on wiring practices provided.

Add to Section 14-2 (c)
Exception: Conductors larger than No. 4 AWG shall be permitted to be terminated directly on the device.

To make provision for larger field wiring connections.

Add Section 14-2(d) and (e):
(d) Conductors or cables between assemblies or components shall have sufficient length, support, and be secured to minimize stress and abrasion on the conductors, cables, bundles and terminations.

Additional clarifications on wiring practices provided.

(e) The cross sectional area of conductors in the wiring duct shall not exceed 40 percent of the cross sectional area of the wiring duct.

The intent is to ensure that there is space available within the wireway to allow for inspection and future additions within the enclosure.

14-3 Machine Wiring.

Add to Section 14-3(a):
Multiple conductor cable shall be permitted to be installed in cable tray in accordance with the requirements of the NEC Article 318.

Reference to the NEC added to clarify the practices frequently used in the automotive industry.

Add to Section 14-3 (c)
All flexible connections shall be provided with a strain relief device.

To minimize damage to cable and connections.

Variation	Rationale

Add Section 14-3(m) thru (q) :

(m) The exposed length of multiconductor cable or flexible conduit to stationary or infrequently moved devices shall not exceed 914 mm (three feet). An additional three feet of cable may be enclosed in a raceway, but in no case shall cable length or its conductors exceed six feet overall. The jacket shall remain on the enclosed portion of any cable.

Maximum length established to limit the excessive use of multiconductor cable or flexible conduit. It is recognized that certain applications may require additional lengths - specific approval will be required.

(n) Power and control circuit conductors shall not be contained in the same multiconductor cable.

(o) To minimize electromagnetic interference, a separate steel raceway shall be used for electronic and other precision signal wiring, including thermocouples, to separate it from all other power and control wiring.

(p) Spare wires shall be provided in all cables and raceways. The spare wires shall run directly from the control panel enclosure to the master terminal box on each station. All spares shall be terminated or stored neatly and bundled together in the bottom of the electrical enclosure with their ends insulated, as indicated by the purchasing division. Spare wires shall be of sufficient length to reach the extreme points of the enclosures unless terminated. Spare wires shall be uniquely labeled "Sp 1," "Sp 2," etc. The number of spares shall be 10 percent of the total wires used or a minimum of two, whichever is greater.

Allows for a reasonable amount of expansion and ease of change which experience has shown to be frequently required in flexible manufacturing installations.

(q) Wiring external to the control panel enclosure shall have a termination at the terminal blocks on the subplate. One wire shall be returned to the main enclosure for test purposes, from a connection between limit switches, pushbuttons or other devices connected in series.

Permits diagnostics to be accomplished from the main enclosure, for all field wiring and field devices.

14-4 Wire Connectors and Connections.

Add Section 14-4(c) thru (e):

(c) Connections should enter the sides or bottom of an enclosure or box.

Added to ensure the environmental protection of the enclosure.

(d) Electrical connections to motors, solenoids and similar devices with integral leads, size No. 4 AWG and smaller, shall be made with ring type pressure connectors (pressure-tool applied). Pressure connectors shall comply with UL 486B. The connectors shall be bolted and insulated with tape which will not support combustion. Soldered or insulation piercing type connectors (lugs) are not acceptable.

Designed for ease of replacement of these devices while ensuring a reliable electrical connection.

Variation

(e) Load side terminations on the main disconnecting means shall be limited to no more than the manufacturer's recommended connections. When required, splitter blocks (single conductor "in" - multiple conductors "out") shall be furnished for additional tap connections for circuit branching (e.g. disconnect switches, fuse blocks and transformer secondaries). A minimum of 10 percent or four spare terminals, whichever is greater, shall be furnished on splitter blocks. Splitter blocks shall be rated for copper conductors. (See 6-4, "Location of Protective Devices".)

Rationale

Provides requirement not covered in NFPA 79 for tap connections. Allows for a reasonable amount of flexibility and ease of change which experience has shown to be frequently required in flexible manufacturing installations.

Variation	Rationale

Chapter 15:   Raceways, Junction Boxes, and Pull Boxes

15-1 General Requirements.

Replace Section 15-1 (b) with:
  (b) Drain holes shall not be permitted in raceways, junction boxes, and pull boxes.

The intent is not to limit manufacturing or design practices, but to assure that the enclosure maintains the intended integrity. Allows for the use of accessories or devices designed to maintain the integrity.

Add Section 15-1(c) ahead of Note
  (c) Prior to installation of wiring, oil and chips shall be removed from raceways and fittings.

15-2 Percent Fill of Raceways.

Replace Section 15-2 with:
The number of conductors permitted in conduits and wireways shall be as shown in Table 15-2. The percentage fill used in these tables is 40 percent. This is the maximum number of conductors permitted in the raceway, including all required spares. Quantities for wire sizes not shown in the table, shall be calculated using 40 percent fill. (The percent fill is equal to the sum total of the cross sectional areas of all the enclosed conductors, divided by the internal, cross sectional area of the raceway.)

Table 15-2 is added for reference and clarity. Wireway fill is 40 percent to allow increased flexibility for field device wiring.

Add Table 15-2

15-3 Rigid Metal Conduit and Fittings.

Add to Section 15-3(a)
  (a) Aluminum, EMT, intermediate, and rigid nonmetallic conduit types shall not be used. All fittings shall be liquidtight. All covers shall be readily accessible. Gaskets shall be of an oil-resistant synthetic material.

Rigid metal conduit is required for increased reliability and crush resistance where subjected to physical damage.

Retitle Exception as Exception No. 1:

Add Section 15-3(a) Exception No. 2:
Exception No. 2: Where metallic raceways are not suitable for environmental conditions other material may be used.

Add Table 15-2: Maximum Capacity for Conduit and Wireway Fill

Conduit 40% Fill: Area of Conductor (sq. in.)

	A	B	C	D																																			

Conductor Size AWG-kcmil

	A	B	C	D	A	B	C	D

(Table column headers: MTW (A), TF/THW/TW (B), TFFN/THHN/THWN (C), XHHW/ZW (D) for each Conduit Trade Size: 1/2, 3/4, 1, 1 1/4, 1 1/2, 2)

Conductor Size AWG-kcmil	MTW (A)	B	C	D
22	0.0066	-	-	-
20	0.0080	0.0088	-	-
18	0.0097	0.0109	0.0062	-
16	0.0119	0.0135	0.0079	-
14	0.0150	0.0206	0.0087	0.0131
14	-	0.0172	0.0117	0.0131
12	0.0191	0.0252	0.0117	0.0167
12	-	0.0222	0.0184	0.0167
10	0.2570	0.0311	0.0184	0.0216
10	-	0.0471	0.0373	0.0216
8	0.0535	0.0598	0.0519	0.0456
8	-	0.0819	0.0845	0.0625
6	0.0908	0.1087	0.0995	0.0845
4	0.1207	0.1263	0.1182	0.0995
3	-	0.1473	0.1590	0.1182
2	0.1713	0.2027	0.1893	0.1590
1	0.2411	0.2367	0.2265	0.1893
1/0	0.2827	0.2781	0.2715	0.2265
2/0	0.3432	0.3288	0.3278	0.2715
3/0	0.3915	0.3904	0.4026	0.3278
4/0	0.4729	0.4877	0.4669	0.4026
250	-	0.5581	0.5307	0.4669
300	-	0.6291	0.5931	0.5307
350	-	0.6969	0.5931	0.5931
400	-	0.8316	0.7163	0.7163
500	-	-	-	-

Wire way 40% Fill: Area of Conductor (sq. in.)

Conductor Size AWG-kcmil	MTW (A)	TF/THW/TW (B)	TFFN/THHN/THWN (C)	XHHW/ZW (D)
22	0.0066	-	-	-
20	0.0080	0.0088	-	-
18	0.0097	0.0109	0.0062	-
16	0.0119	0.0135	0.0079	-
14	0.0150	0.0206	0.0087	0.0131
14	-	0.0172	0.0117	0.0131
12	0.0191	0.0252	0.0117	0.0167
12	-	0.0222	0.0184	0.0167
10	0.2570	0.0311	0.0184	0.0216
10	-	0.0471	0.0373	0.0216
8	0.0535	0.0598	0.0519	0.0456
8	-	0.0819	0.0845	0.0625
6	0.0908	0.1087	0.0995	0.0845
4	0.1207	0.1263	0.1182	0.0995
3	-	0.1473	0.1590	0.1182
2	0.1713	0.2027	0.1893	0.1590
1	0.2411	0.2367	0.2265	0.1893
1/0	0.2827	0.2781	0.2715	0.2265
2/0	0.3432	0.3288	0.3278	0.2715
3/0	0.3915	0.3904	0.4026	0.3278
4/0	0.4729	0.4877	0.4669	0.4026
250	-	0.5581	0.5307	0.4669
300	-	0.6291	0.5931	0.5307
350	-	0.6969	0.5931	0.5931
400	-	0.8316	0.7163	0.7163
500	-	-	-	-

(Wireway Trade Sizes: 2.5 X 2.5, 4 X 4, 6 X 6, 8 X 8, 12 X 6 each with columns A, B, C, D)

Variation   **Rationale**

Replace Section 15-3(d) with:

(d) All fittings shall be threaded per ANSI B.11.20.1 General Pipe Threads.

Threadless fittings are not allowed to provide additional mechanical integrity of the system.

Replace Section 15-3(f) with:

(f) Where conduit enters a box or enclosure, a bushing or fitting providing a smoothly rounded insulating surface shall be installed to protect the conductors from abrasion unless the design of the box or enclosure is such as to afford equivalent protection. A locknut shall be provided both inside and outside the enclosure to which the conduit is attached. Fittings shall maintain the NEMA rating of the enclosure. Conduit shall be installed so that liquids tend to run off the surface instead of draining towards the fittings. Conduit should enter the side or bottom of enclosures.

Clarification: requires the use of locknuts for terminating conduits in a control enclosure. Added to ensure the environmental protection of the enclosure.

Exception: Where threaded hubs or bosses that are an integral part of an enclosure provide a smoothly rounded or flared entry for conductors, the bushing inside may be omitted.

15-4 Intermediate Metal Conduit.

Delete Section 15-4 and leave blank intentionally.

Section 15-4 does not apply per requirements of Section 15-3(a).

15-5 Electrical Metallic Tubing.

Delete Section 15-5 and leave blank intentionally.

Section 15-5 does not apply per requirements of Section 15-3(a).

15-6 Schedule 80 Rigid Nonmetallic Conduit.

Delete Section 15-6 and leave blank intentionally.

Section 15-6 does not apply per requirements of Section 15-3(a).

Variation	Rationale

15-7 Liquidtight Flexible Metal Conduit and Fittings.

Add to Section 15-7(b):

Connectors for liquidtight flexible metal conduit shall be liquidtight, made of metal, and meeting the requirements of UL 514. Fittings shall have sufficient thread length to accommodate a gasket assembly, a box wall thickness of 0.125 inch and a locknut and bushing. Connectors shall be "union" type.

Additional requirement to ensure Type 12 enclosure integrity. Union type connectors required for serviceability.

15-9 Flexible Metal (Nonliquidtight) Conduit and Fittings.

Replace Sections 15-9(a) through (c) with:

Flexible metal, nonliquidtight conduit shall not be used, except for thermocouples and similar devices.

Flexible metal nonliquidtight conduit applies only for the exception.

15-10 Wireways.

Replace Section 15-10(d) with:

(d) A lay-in wireway shall be furnished between electrical enclosures when the number of conductors exceeds that for which a 2-1/2 inch conduit is suitable. (2 square in. wire cross-sectional area).

Added for serviceability and accessibility for additional wiring.

Replace Section 15-10(e) with:

(e) Wireways shall be Type 12 minimum, without knockouts.

To maintain the Type 12 integrity. To insure no unused openings. Knockouts can leak.

Add Section 15-10(f):

(f) All covers shall be capable of opening at least 90 degrees.

Access required for serviceability.

Add Section 15-10(g):

(g) Wireways shall be fitted with telescopic sections in the horizontal and vertical portions between the enclosure and equipment to provide for variations up to plus or minus five inches.

To accommodate variation at time of installation.

Add Section 15-10(h):

(h) Corners, bends, edges, etc., shall have all burrs removed. Additional protection shall be provided to protect conductor insulation at all sharp bends and drop points. Such protection may consist of fiber, plastic or other material to cover the edge or corner with sufficient radius to prevent damage to the insulation.

To insure the wire insulation is protected.

Variation	Rationale

15-12 Junction and Pull Boxes.

Add Section 15-12(a) thru (h):

(a) Boxes shall be constructed in conformance with applicable sections of this standard and with either of the following:

    1. NEMA ICS6 and 250, Types 12 and 13, Industrial Use.

    2. UL 508 and UL 50, Types 12 and 13, Industrial Control Equipment.

The original NFPA text "constructed to exclude" was expanded to recognize industry standard Type 12 and 13 environmental protection.

(b) Wall thickness of steel boxes shall meet the requirements of UL 50.

Added to reference an industry standard.

(c) Boxes with an access opening wider than 102 mm (4 in.) shall include provisions for mounting a subplate. The height and width of the door opening shall be greater than the corresponding height and width of the subplate to be enclosed.

Additional text to require that the box is large enough to accommodate a panel should have panels provisions included. The panel should be sized such that installation and removal of the panel can be accomplished easily.

(d) Subplates shall be a minimum of 14 gauge (2 mm) for subplates (200 in²) 1290 cm² and smaller and 12 gauge (2.7 mm) for subplates larger than (200 in²) 1290 cm² for mounting of components by drilling and tapping the subplate. Maximum screw size is No. 10 for 14 gauge and ¼" for 12 gauge. Additional reinforcement or heavier gage subplates shall be provided where larger screws are required.

Subplate gauge requirements are required to ensure minimum of 2 threads engagement for drilled and tapped subplates.

(e) Any holes or cutouts through the enclosure walls shall be properly sealed to maintain the original protection category of the enclosure. Mounting feet or other suitable means external to the enclosure shall be provided for mounting.

Type 12 rated conduit hubs and other hole closing devices should be used to maintain the original type 12 and 13 protection. The mounting feet should be external to the. equipment cavity to maintain the original protection.

(f) Covers larger than 193 cm² (30 square inches) shall be hinged. Covers with hinges shall allow the cover to be easily removed with a tool. Hinges shall allow the cover to open at least 120 degrees.

Large covers shall be hinged and must open 120° for improved serviceability.

(g) Cover fasteners shall secure the cover in the closed position and require a tool to operate. Cover fasteners shall be captive to either the box or cover. Fasteners should be the quick release type. Vault type hardware shall be used on boxes with hinged doors when the access opening is 1016 mm (40 in.) high or more.

Cover fasteners shall be captive to the enclosure to ensure they don't become misplaced. Large doors require vault type hardware to minimize the number of cover fasteners needed to seal the door.

Variation	Rationale

(h) Cover sealing gaskets shall be at least 3.2 mm (.125 in.) thick and shall be held firmly and permanently in place with oil-resistant adhesive or equivalent. The gasket shall be of an oil-resistant material which meets UL 508 Type 12 and 13 requirements.

Requirement for oil resistant adhesive is added to ensure compatibility. Equivalent gasket types like foamed-in place gaskets may be acceptable. The thickness requirement is to insure a minimum level of quality for enclosure gaskets. Gasket properties are defined by referencing UL 508.

Variation	Rationale

Chapter 16:   Motors and Motor Compartments

16-2 Mounting Arrangement.

Add Section 16-2(e):
(e) An adjustable base or other means shall be provided when belt or chain drives are used.

Required to permit proper tensioning of the belts or chains.

Add Section 16-2(f):
(f) Where failure of any motor mounting can create a safety hazard, the motor shall be safety cabled to a permanent rigid part of the structure.

Requirements were added to address concerns in the automotive industry.

16-3 Direction Arrow.
Add to Section 16-3:
A direction arrow shall be provided for all non-reversing motor applications.

Experience has indicated that directional arrows are useful for all motors.

16-4 Motor Compartments.

Add to Section 16-4:
The motor compartment or mounting space shall be of sufficient size to accommodate a NEMA frame horsepower one size larger, at the same speed, than that recommended by the industrial equipment supplier.

Added requirement provides flexibility for processing changes.

16-5 Marking on Motors.
Add Section 16-5(a):
(a) Motor nameplates shall also include, locked-rotor current, maximum KVAR switched with the motor, and NEMA Design Type.

NEMA Design Type and locked rotor current are required for proper selection of motor controller and disconnecting devices.

Add Section 16-5(b):
(b) Special characteristics of AC motors shall be shown on a separate nameplate mounted adjacent to the conventional motor nameplate unless already indicated on the standard nameplate. The manufacturer's catalog numbers shall not be considered sufficient to designate such characteristics.
 Typical examples are:
  (1) Special insulation.
  (2) Special shaft length.
  (3) Special torque.
  (4) Special balance.
  (5) Special lubrication.
  (6) Special mounting.

Special motors cause added maintenance concerns. Nameplates will help ensure correct servicing or replacement.

Variation   Rationale

Add Section 16-6:
16-6 Motor Applications, General.

Requirements in this section address many of the concerns in the automotive industry. The correct application of motors requires careful specification that is not addressed by NFPA 79. The SAE specification is based on experience within the automotive industry.

Add Section 16-6(a) thru (i):

(a) Electric motors shall be in compliance with the motor specifications indicated on the "Equipment Data Form".

(b) "Special feature" motors shall not be used unless specific approval has been received from the specifying authority. Where special application dictates a motor design other than standard NEMA foot mounting, approval in writing from the purchasing division shall be required at time of quotation. It shall be the responsibility of the machinery or equipment manufacturer to consult with the motor manufacturer on applications where high radial bearing loads are involved.

(c) Motors located in "Hazardous Locations" shall meet all the requirements of this specification, except that they shall be "Explosion-Proof" for the Class, Group, Division and Code of service, as defined by the National Electrical Code and approved by Underwriter's Laboratories to suit the hazardous location involved.

(d) All motors shall be sized to maximize motor efficiency and applied so that the rms brake horsepower load connected to the motor shaft shall not be more than 100 percent of the motor nameplate brake horsepower rating during normal operation of the equipment. In no case shall the intermittent peak horsepower exceed 160 percent of nameplate brake horsepower.

(e) All motors shall be applied so that the connected load can be started and accelerated at 90 percent of rated voltage.

(f) Each AC induction motor shall be sized to have a minimum connected load of 75 percent of the full load horsepower rating of the motor.

(g) All motors for belt driven fan applications shall meet all requirements of this specification and the acceleration time shall not exceed 10 seconds. The formula for determining acceleration time shall be as follows and is based on 150 percent average accelerating torque.

$$\text{Acceleration in Seconds} = \frac{\text{Actual Fan WK}^2 \text{ in lb.-ft}^2. \text{ x Fan Speed in rpm}^2}{2.49 \times 10^6 \times \text{Motor Horsepower}}$$

If the calculated fan load horsepower falls between standard motor horsepower ratings, the higher horsepower motor shall be used. This established motor horsepower, when used in the acceleration in seconds formula, shall not be arbitrarily increased to reduce the acceleration time. The only acceptable method of decreasing the acceleration time shall be by varying the fan WK^2 and/or the fan speed in rpm.

(h) The operating cycle for multi-speed and single-speed duty-cycle motors must be completely defined before it can be determined whether the specific motor will handle the operation.

Duty-Cycle Operation: The following information should be used as a guide where duty-cycle or rapid reversal motors are required:

(1) A reversal at no load is equivalent to one acceleration plus on plug stop.

(2) One application of DC dynamic braking is equivalent to one acceleration.

(3) The losses in the rotor due to a plug stop are three times those due to an acceleration.

(4) The number of reversals that can be obtained with an externally connected WK^2 is equal to:

$$\frac{\text{(Int.) WK}^2}{\text{(Int.) WK}^2 + \text{(Ext.) WK}^2} \text{. X Reversals/Minute @ no load}$$

Where: Int. is Rotor WK^2. Ext. is connected WK^2 referred back to the motor shaft.

(5) If the motor is braked to a stop, the number of starts and stops possible without overheating is approximately four times that possible with plug stopping. The brake life capabilities must be considered when using a brake.

(i) For applications requiring two or more motor reversals per minute, electric motors with high torque - high slip characteristics (5 to 8 percent), and having at least class "B" insulation, shall be furnished.

Chapter 17   Ground Circuits and Equipment Grounding

17-1 General.

Add Figure 17-1
Reference Figure 17-1.

<u>Fig. 17-1</u>

Grounding Symbols

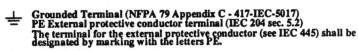 Grounded Terminal (NFPA 79 Appendix C - 417-IEC-5017)
PE External protective conductor terminal (IEC 204 sec. 5.2)
The terminal for the external protective conductor (see IEC 445) shall be
designated by marking with the letters PE.

G 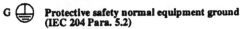 Protective safety normal equipment ground
(IEC 204 Para. 5.2)

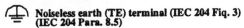 Noiseless earth (TE) terminal (IEC 204 Fig. 3)
(IEC 204 Para. 8.5)

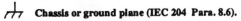 Chassis or ground plane (IEC 204 Para. 8.6).

N   Grounded circuit (neutral) conductor
(N must not be re-grounded
on load side of transformer)

17-2 Grounding Conductors.

Add to Section 17-2(d):
This permissible use of the machine members and structural parts of the electrical
equipment is to accomplish bonding of the structures and is not a substitute for the
equipment grounding requirements of Section 17-3. Ground continuity is
illustrated in Figure 17-9A thru 17-9F.

*The use of machine members
or structural parts for the
grounding circuit provides a
high impedance as well as an
unreliable and unpredictable
ground return path due to
variations in materials,
bearings, painted surfaces,
structural connections, etc.
The text in Section 17-3
requires the inclusion of a
grounding conductor.*

17-3 Equipment Grounding.

Replace Section 17-3 with:
All exposed, noncurrent-carrying metal parts of equipment, such as control
enclosures, raceways, control stations, separately mounted apparatus, and portable
and pendant accessories, which are likely to become energized, shall be effectively
grounded with an equipment grounding conductor. Wiring to all electrical devices,
except those which are identified as double-insulated, shall include an equipment
grounding conductor. Reference Figure 17-3A and 17-3B.

*Section 17-3 was revised and
Figures 17-3 A and B were
added to more specifically
detail the equipment
grounding requirements.*

Variation   Rationale

Add Section 17-3(a) thru (d):

(a)   Equipment grounding conductors shall be terminated in each electrical enclosure on a common ground bar which is bonded to the subplate. A grounding type terminal strip, approved for the use, is acceptable. A conventional terminal strip with individual terminals jumpered together is not acceptable.

Fig. 17-3A

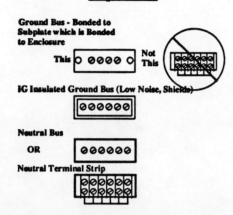

(b) Insulated or dedicated equipment grounding conductors shall be permitted to be used for sensitive computer/electronic equipment provided that the requirements of the National Electrical Code, Article 250, are complied with. The isolated equipment grounding conductor shall be connected to ground at the equipment grounding terminal described in Section 17-5.

(c) Ground rods installed at equipment shall be bonded to the equipment grounding terminal described in Section 17-5.

Variation   Rationale

(d)   The shield for shielded cables may be terminated separately from the equipment   grounding   conductors.

Fig. 17-3B

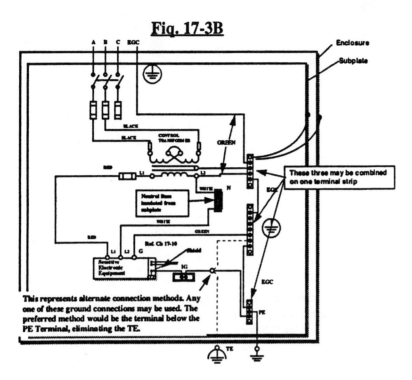

17-7 Control Circuits.

Add Section 17-7(a):

(a)   The specifying authority shall indicate on the Equipment Data Form in Appendix G whether circuits shall be grounded or ungrounded. Where control circuits are required by applicable laws or codes to be grounded, the supplier shall notify the specifying authority of the requirement and design the control circuit to conform. Reference Figure 17-7A.

Added to allow this documentation to serve as a purchasing specification.

Fig. 17-7A

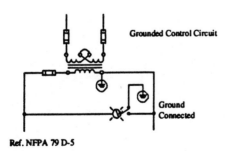

Variation

Add Section 17-7(b):

(b) Ungrounded control circuits shall be equipped with ground detector lights or other fault detection device. Reference Figure 17-7B.

Added to comply with accepted practice in the automotive industry.

<u>**Fig. 17-7B**</u>

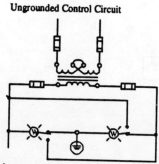

Ungrounded Control Circuit

Add Section 17-7(c):

(c) The grounding of the control circuit shall be accomplished by direct connection from the transformer or power supply, to the equipment ground bus in the enclosure. The grounded circuit conductor shall not be connected to ground at any other point. Reference Figure 17-3B.

Added to comply with accepted practice in the automotive industry

17-8 Lighting Circuits.

Add exception to 17-8(a):

Exception: When machine lighting and maintenance lighting circuits are supplied from the equipment control panel, and remain energized with the disconnect in the off position, the conductors to the lighting fixture shall be yellow.

Reference Figure 17-8.

Added to call attention to the requirement that wires powered by sources not turned off by the main disconnect must be yellow.

<u>**Fig. 17-8**</u>

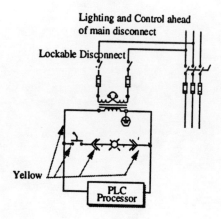

Lighting and Control ahead of main disconnect

Lockable Disconnect

Yellow

PLC Processor

Variation

Rationale

Replace Section 17-8(b) with:

(b) Where the lighting circuit is supplied by a separate isolation transformer, the grounding shall occur at the transformer. Where the equipment maintenance lighting circuit is supplied directly from the plant lighting circuit, no additional grounding shall be permitted.

Revised to avoid confusion that could lead to multiple grounds on the lighting circuit neutral wire conductor.

17-9 Grounding of Power Circuits.

Add Section 17-9:

17-9 Grounding of Power Circuits.
Each power circuit shall be installed with an equipment grounding conductor to provide a continuous ground connection between equipment and the grounding electrode conductor. The raceway system, cable trays, and cable sheaths shall not be permitted to serve as equipment grounding conductors. Install insulated copper ground wires with power circuits in conduits, ducts, and cable trays. Provide armored cable assemblies with bare copper ground wires in the interstices. Provide green insulated copper ground wires in type TC tray cables and flexible cords for power circuits. Reference Figures 17-9A thru 17-9F.

Personnel safety, equipment protection, and system performance are the basic objectives of acceptable grounding practices. This supplement may exceed various code minimum requirements in order to meet the objectives and was added to ensure electrical continuity between equipment enclosures and the facility service ground (or separately derived ac system), without relying on raceway connections.

Fig. 17-9A

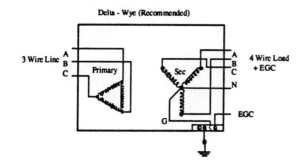

Fig. 17-9B

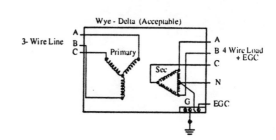

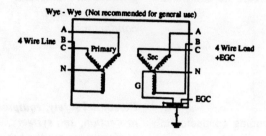

Fig. 17-9C

Wye - Wye (Not recommended for general use)

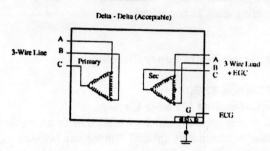

Fig. 17-9D

Delta - Delta (Acceptable)

Fig. 17-9E

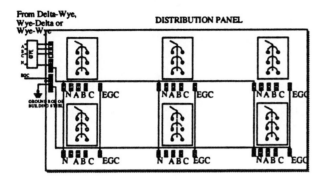

Fig. 17-9F

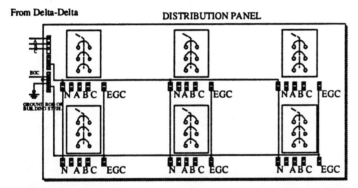

Add Section 17-10:

17-10 Shielding and Grounding. Equipment chassis shall be interconnected with short, wide straps to the power system ground of 17-9, forming an equipotential plane. Shields and/or sheaths of each control, signal, and communications circuits shall be grounded at the signal source-end and grounded via a transient voltage surge suppression (TVSS) device at the opposite end. Signal circuits routed between noncontiguous areas shall be installed in metallic raceways bonded at couplings and bonded to ground at intervals equaling 1/8 the wavelength of the predominant signal frequency. Reference Figure 17-3B.

Personnel safety, signal reference integrity, and transient protection are the basic objectives of acceptable RF shielding and grounding practice. This requirement was added to promote consistent grounding methods for instrumentation circuits and to minimize ground loops.

Chapter 18   Electronic Equipment

18-1 General.
Add Section 18-1(a):
(a) Electronic devices shall conform to ANSI/NEMA ICS-1 through ICS-4 and ICS-6.

18-2 Basic Requirements.
Replace Section 18-2(b) with:
(b) Subassemblies shall be readily removable for inspection or replacement unless third party certification would be voided, and shall be labeled with the manufacturer's name and part number.

This is to protect third party proprietary hardware warranty. Manufacturer is to provide identification on subassemblies that are removable.

Replace Section 18-2(d) with :
(d) Electronic devices requiring power for the purpose of sustaining memory shall be supplied with battery back-up of sufficient capacity to prevent memory loss for a period of at least fourteen (14) days. Batteries shall be safely replaceable while under power and shall not disrupt the operation of the system.

Technology exists that provides fourteen (14) days back-up. Facilities may occasionally shut down for fourteen (14) days.

18-3 Programmable Electronic Systems.

Add Section 18-3(c) thru (h):
(c) Power supplies to electrical devices shall be regulated type incorporating transient suppression and isolation from electrical noise. These power supplies shall be sized to permit a 25 percent increase in the load. In addition, spare I/O slots shall be provided for expansion of a minimum of 20 percent inputs and 20 percent outputs above the system's initial requirements. The programmable device shall contain 20 percent excess memory (50 percent shall be contiguous) above the system's initial requirements. There shall be 20 percent spare of each item listed in the executive table (timers, counters, logic lines, etc.).

The spare requirements are to provide for engineering modifications and continuous improvement of equipment.

(d) The programmable devices shall permit loading, recording, and verifying programs with the main programming device.

(e) The PES shall not increment/decrement counters nor otherwise cause false movement or change in retained data when power is turned off or on.

This is to prevent incorrect machine motion and incorrect machine / control / production status.

(f) Status indicating lights on the CPU shall verify normal operation. Typical indications are: AC and DC power, processor cycling, parity error, memory fault, memory support battery low, processor fault, power supply fault, and I/O fault.

These status indicating lights are to aid in system troubleshooting

Variation	Rationale

(g) Error Detection and Fault Isolation (EDIF) shall be integral and shall localize faults to the lowest modular (plug-in) level.

The addition of these requirements are to aid in trouble shooting machine control.

(h) Local fault annunciation via lights, and program accessible status bits shall be available.

Add Section 18-4:
18-4 Programming and Data Access.

Add Section 18-4(a)and (b):
(a) The use of the program panel interface shall not interfere with the data communications interfacing.

(b) The main programming device shall indicate during programming when an output address has been previously used.

Add Section 18-5:
18-5 Electronic Drives.

Add Section 18-6(a) thru (d):
(a) Electronic drive systems shall be designed to monitor parameters, to prevent damage to the drive or equipment and to maintain its safe operation. Fault conditions shall be indicated for diagnostic purposes.

(b) Electronic drive systems shall be modular in construction for ease of maintenance and replacement.

(c) Where load conditions or reduced speeds can cause motor overheating, thermal detectors within the motor winding shall be provided. Thermal detector(s) shall be interlocked with the electronic drive system.

(d) Electronic drives shall include an electromechanical contactor on the load side if electrically interlocked with the drive) or on the line side that will remove power from the ungrounded motor conductors. This contactor may be installed as an addition to the drive or may be an integral part of the drive control.

To ensure that power is removed during emergency stop conditions. The exception allows elimination of this requirement.

Exception: Where a drive failure does not cause a hazard to personnel .

Appendix A   Glossary of Terms

Blanking: The option of intentionally creating a gap in the sensing field to accommodate machine operations such as stock feeding.

Replace "Control Enclosure" with:

Control Enclosure: A control enclosure is defined as an enclosure with panel(s) (subplates) which house electrical power and control equipment such as disconnect switches, relays, starters, contactors, and other related control equipment or electronic control and logic equipment such as programmable controllers, microprocessors, computers, CRT's and other related electronic devices or a combination thereof.

A small enclosure is one which accommodates a subplate having 9678 sq. centimeters (1,500 sq. in.) or less of area.

A large enclosure is one which accommodates a subplate having more than 9678 sq. centimeters (1,500 sq. in.).

Control Panel: See Panel (Subplate)

Junction Box: A box with a blank cover that serves the purpose of joining different runs of raceway or cable and provides space for the connection and branching of the enclosed conductors. (ANSI/IEEE Std 100-1984)

Light Color: A light color is a color with a reflectance greater than Y(%) = 78 CIE specification (i.e., white as defined by ANSI Z53.1 has a Y% value range of 78% to 90%).

Muting: A means of bypassing the protective function of a safeguarding device.

Object Sensitivity: The minimum object size that will be detected anywhere in a sensing field regardless of the plane of intrusion.

Replace "Panel" with:

Panel (subplate): See Subplate

Point of Operation: The area of the hazardous machine where material is actually positioned and work is being performed during any process such as shearing, punching, forming, welding or assembling.

PSDI: Presence Sensing Device Initiation: The common term used in an industrial machine control application in which a presence sensing device is used to start the cycle of a machine (i.e. no start button is used).

PSSD: Presence Sensing Safety Device: a device designed, constructed and arranged to create a sensing field, area or plane (usually optical) that will detect the presence of personnel for the purpose of stopping hazardous operation before personnel are able to reach the danger point, thus preventing injury.

Variation	Rationale

Pull Box: A box with a blank cover that serves the purpose of joining different runs of raceway or cable and provides space for the enclosed conductors.
rotatable handle and an internal latching system. The internal linkage is actuated by the rotating handle and simultaneously releases or closes latches at top and bottom (2 points) or top, bottom and middle (3 points) of door.

Pushbutton Box: An enclosure used to mount assemblies of pushbutton switches.
An element of an electric controller consisting of a slab or plate on which various component parts of the controller are mounted and wired.

Rated Conditional Short Circuit Current
The rated conditional short-circuit current of an equipment is the value of prospective current, stated by the manufacturer, which the equipment, protected by a short circuit protective device specified by the manufacturer, can withstand satisfactorily for the operating time of this device under the test conditions specified in the relevant product standard.
The details of the specified short-circuit protective device test conditions shall be made available by the manufacturer.

Reference IEC 947-1 (1) 4.3.6.4

Note 1 For ac., the rated conditional short-circuit current is expressed by the r.m.s. value of the ac. component.
2 The short-circuit protective device may either form an integral part of the equipment or be a separate unit.

Safety Distance: The distance between the pinch point or point-of-operation and the presence sensing safety device sensing field which ensures that the operator cannot reach the danger point before the machine comes to a full stop.

Sub Assembly: An assembly of electrical devices connected together which forms a simple functional unit in itself.

Replace "Subpanel" with:
Subpanel: See Sub Assembly.

Subplate: An element of an electric controller consisting of a slab or plate on which various component parts of the controller are mounted and wired.

Supplementary Protection: A device designed to open the circuit at a predetermined value of time versus current within electrical equipment where branch circuit overcurrent protection is already provided, or is not required.

Terminal Box: A box with a blank cover that serves the purpose of joining different runs of raceway or cable and provides space for the connection and branching of the enclosed conductors on terminals.

Vault Type Hardware: A door fastening means which consists of an external rotatable handle and an internal latching system. The internal linkage is actuated by the rotating handle and simultaneously releases or closes latches at top and bottom (2 points) or top, bottom and middle (3 points) of door.

EQUIPMENT DATA FORM

The following information is provided by the end user of the equipment to ensure proper design, application, and utilization of electrical equipment. Default conditions are shown in parenthesis, special requirements will be noted in the adjacent space.

DATE			PURCHASE ORDER #	
EQUIPMENT ASSET #			EQUIPMENT SERIAL #	
INQUIRY #		FOR USE AT		
USER DRAWING #				
MFG/SUPPLIER	NAME		ADDRESS	
	PHONE/FAX #		CONTACT	
END USER	NAME		ADDRESS	
	PHONE/FAX #		CONTACT	
DESCRIPTION OF EQUIPMENT				

STANDARDS or CODES APPLICABLE TO THIS EQUIPMENT: SAE (*Standard*) ☐   Other ☐

_____

Third party certification required: Yes ☐   No ☐ Who

_____

Are there deviations to the applicable Standards or Codes? Yes ☐   No ☐ (If yes,

document using Deviation Request Form)

For items 1 - 5 refer to Chapter 3 - General Operating Conditions

1.   Ambient temperature range (5 to 40° C) _____

2.   Altitude (3300) ft.   _____

3.   Humidity (20 - 95% non condensing)   _____

4.   Supply Voltage

       Electrical supply required/available: (Nominal Systems Voltage)

       ☐ 600 Vac, 3 phase, 3 wire   ☐ 480/277 VAC, 3 phase, 4 wire

       ☐ 480 VAC, 3 phase, 3 wire   ☐ 240 VAC, 3 phase, 3 wire

       ☐ 600 Vac, 3 phase, 3 wire   ☐ 208/120 VAC, 3 phase, 4 wire

       ☐ 480 VAC, 3 phase, 3 wire   ☐ OTHER_____

Frequency: 60 Hz ☐ 50 Hz ☐, ☐_____Hz

Maximum horsepower for motors permitted to be started directly across the incoming

supply lines: _____Hp

Voltage Supply Requirements:

(a) Voltage (90-110% of rated voltage) _____

(b) Frequency (+/-2% of rated frequency) _____

(c) Harmonic Distortion (up to 10% of total rms sum of the 3rd through 7th harmonics.

Up to an additional 2% rms of 7th through 30th harmonics. _____

(d) Radio Frequency (2% RMS above 10 KHz voltages) _____

(e) Impulse Voltage (200% peak voltage up to 1 ms duration with a rise time of 500 ns

to 500 us.) _____

(f) Voltage Drop (Reduction of 50% of peak voltage for 1/2 cycle or 20% for 1 cycle.

More than 1 second between successive reductions.) _____

(g) Micro-Interruption (Supply disconnected or at zero voltage for 3 milisec. at any

random time in the cycle. (More than 1 second between successive reductions.)

_____

(h) Voltage Imbalance (The voltage imbalance shall be as specified in ANSI C84.1)

_____

5a.   Environmental conditions such as corrosive atmospheres, particulate matter, etc.

_____

5b.   Control Enclosure Construction (if other than steel):

☐ Aluminum

☐ Iron

☐ Stainless Steel

☐ OTHER _____

5c.   NEMA type if other than type 12:

_____

5d.   Subplate Finish: ☐ Gloss White and/or ☐ Gloss Orange

6. Radiation _____

7. Vibration, shock _____

8. Indicate possible future changes in the equipment that will require an increase in the electrical supply requirements, spare space, etc._____

   _____

   _____

   _____

9.

10. (a) Motor utilization voltage rating: 550 VAC ☐ 460 VAC ☐ 230 VAC ☐ 200 VAC ☐ 115 VAC ☐ OTHER _____

    (b) Phase Loss Protection Required: ☐ Yes (for motor ____Hp & above) No ☐, or as noted.

    _____

    (c) Motor starters provided with Type 2 protection: ☐ Yes ☐ No

    (d) Dual windings: 460/230 VAC _____

11. Power Factor Correction and Harmonic Frequency Control:

    ☐ Power Factor per Section 16   ☐ Harmonic Frequency Control per Sect. 16

12. Control Circuit: grounded ☐ ungrounded ☐

    Special grounded/ungrounded circuit required: Yes ☐ No ☐

13. Type of main disconnecting device to be provided:

    ☐ Circuit breaker; ☐ Fusible (class of fuses ____);

    Interrupting Capacity of main disconnect _____

14. Technical documentation:

    Media _____

    Language _____

15. Size, location, and purpose of ducts, open cable trays, or cable supports to be provided by the user (additional sheets to be provided where necessary).

_____

_____

_____

16. Special limitations on the size or weight which may affect the shipping of the equipment

or control assemblies to the installation site.

Maximum dimensions: height _____ length _____ width _____

maximum weight _____

17. The expected cycles of operation will be: _____ per hour.

Expected cycle life: _____years

18. Length of time it is expected that this maximum rate or repetition will be repeated without

stopping: _____ (min) (hr)

19. Approved materials and components to be selected from:

_____

_____

Other: _____

_____

_____

_____

_____

20a. Task lighting required: ☐ Yes, ☐ No

20b. Control Enclosure lighting: ☐ Yes, ☐ No

Appendix H

Application Guide for Power Factor
Correction and Harmonic Filtering

1   **Scope**-The application of electronic power converter equipment in industrial machinery creates harmonic voltages and currents that sometimes adversely affect power factor correction capacitors and other electrical loads. This application guide is intended to provide guidance in the selection and application of power factor correction capacitors and harmonic filters.

1.1   **Purpose**-This document provides a proactive approach to power factor correction and harmonic filtering, to be utilized by those involved in the use or supply of industrial machinery and equipment connected to three phase low voltage (600 volts or less) industrial power systems.

2   **References**

2.1   The following related publications are available as supplemental information:

ANSI/NFPA 70 (1993), National Electrical Code (NEC) Article 460, Section A, Capacitors rated 600 volts and under.

NEMA Standard CP-1, Shunt Capacitors

IEEE Standard 18-1992, Shunt Power Capacitors

UL508A Standard, Industrial Control Equipment

NEMA Standard MG 1-1987, Motors and Generators

ANSI/IEEE Standard 141-1986, IEEE Recommended Practice for Electric Power Distribution for Industrial Plants (a.k.a. "IEEE Red Book")

IEEE Std 519-1992, Recommended Practices and Requirements for Harmonic Control in Electrical Power Systems

2.2   **Definitions**

Adjustable speed drive: An electronic power converter that varies the speed of a motor.

Electronic power converter: An electronic device that converts voltage, frequency, or current. Examples are: adjustable speed drives, welding power supplies, uninterruptible power supplies, battery chargers, lighting ballasts.

Harmonic: A sinusoidal component of a periodic wave or quantity having a frequency that is an integral multiple of the fundamental frequency.

Motor control center: An assemblage of motor controllers and related devices in a common enclosure.

Non-linear load: A load that draws a non-sinusoidal current wave when supplied by a sinusoidal voltage source.

Plug-in busway: A metal enclosed structure containing rigid conductors and tap points, used to distribute power throughout a manufacturing area.

Sparing scheme: A power distribution system concept that uses one transformer to provide backup capability for several other transformers. Used in lieu of double ended substations, where one transformer backs up only one other transformer.

3   Power Factor Correction

3.1   Application Warnings

3.1.1   Corrective kVAR exceeding the value required to produce unity power factor at the motor may cause overexcitation resulting in high transient voltages, currents, and torques that can increase safety hazards to personnel and can cause possible damage to the driven or driving equipment.

3.1.2   Redundant schemes, backup schemes, sparing schemes, double ended substations, networks, and loop distribution systems require careful consideration of the effects of switching and a procedure to switch without overcorrecting power factor.

3.1.3   When power factor correction capacitors are switched with an individual motor, the maximum value of corrective kVAR must not exceed the value required to raise the no load displacement (60 hertz) power factor to unity.

3.1.4   When power factor correction capacitors are not switched with motors, but instead are connected at a secondary substation transformer, the maximum kVAR should not exceed 25% of transformer kVA rating.

3.1.5   For applications where overspeed of the machine is contemplated (i.e., induction generators, paralleled centrifugal pumps without check valves) the maximum corrective kVAR should be further reduced by an amount corresponding to the square of the expected overspeed. (From NEMA MG 1-1987, Section 20.86)

3.2   Power Factor Correction Philosophy

3.2.1   For economic reasons, it is generally better to improve power factor for multiple motors rated 20 HP and less on a load area basis as a part of the overall plant power distribution system.

3.2.2   For safety reasons and to avoid overvoltage conditions, it is generally better to switch power factor correction capacitors with individual motors rated more than 20 HP.

3.3   Power factor correction capacitors should not be switched with individual motors where:

3.3.1   The application uses a single motor for reversing, plugging, jogging, inching, or multi-speed operation.

3.3.2   The application uses a clutch, brake, solid state rectifier, open transition wye-delta reduced voltage starter, or DC injection braking.

3.3.3   Motors that are used on an intermittent basis.

3.3.4   Motors that are routinely restarted while still turning.

3.3.5   Motors that are used in regenerative applications where the load can drive the motor, such as cranes, hoists, or elevators.

3.3.6   Motors that are controlled by adjustable speed drives or solid state (soft) starters.

3.4   Motors rated greater than 20 HP should be provided with power factor correction capacitors rated by one of the following methods:

3.4.1   The kVAR rating applied shall be rounded down from the motor manufacturer's maximum recommendation (there may be substantial differences between different manufacturers).

3.4.2   The kVAR rating applied shall be rounded down from:

kVAR = Volts x Motor Amps (no load) x 1.732/1000 (Eq. #1)

3.5   Each secondary substation should include power factor correction capacitors rated to correct the aggregate of all motors rated 20 HP or less, subject to the maximum of 25% of transformer rating mentioned in Section A.3.1.4.

3.6   Capacitor Installation

3.6.1   Capacitors should not be located directly above heat generating devices.

3.6.2   Capacitors that are switched with motors should be connected between the motor starter contactor and the thermal overload relay.

3.6.3   Capacitors connected to motors on reduced voltage starters should include a separate contactor to energize capacitors only after motor is at full voltage.

3.7   **Capacitor Application**-See Figure 1 for alternative locations for power factor correction capacitors. The relative merits and drawbacks for each are discussed below.

3.7.1   Generally, power factor will be improved, transformer and conductor capacity will be freed up, and current, voltage drop, and I^2R losses will be reduced from the connection point of the capacitors in the upstream direction (towards the line side). These benefits will not occur in the downstream (load side) direction.

3.7.2   Generally, capacitors not switched with loads can cause overcorrection during conditions of light loading.

3.7.3   Generally, the further upstream that power factor correction is accomplished, the less will be the cost of the installation, inspection, service, and maintenance.

4 Harmonic Filtering

4.1 Application Warnings

4.1.1 Harmonic filters connected to harmonic generating loads in plants with power factor correction capacitors on the power distribution system should be magnetically isolated from the power factor correction capacitors. See Section A.4.4.

4.1.2 Redundant schemes, backup schemes, sparing schemes, double ended substations, networks, and loop distribution systems require careful consideration of the effects of switching and a procedure to switch without creating resonance.

4.1.3 Harmonic filters provide power factor correction as well as harmonic filtering. The maximum value of corrective kVAR must not exceed the value required to raise the no load power factor to unity.

4.2 When to investigate

Harmonic voltages and currents exist in all power systems. They are generated by many types of electronic switching devices, including adjustable speed drives, switching power supplies, uninterruptible power supplies, lighting ballasts, and other non-linear devices. Generally, they do not cause problems or need correction unless the total harmonic current exceeds 10% of the fundamental current or 10% of transformer rated current, whichever is less.

4.3 Harmonic Filtering Philosophy

4.3.1 For power system performance reasons, it is generally better to correct for harmonic currents on an individual machine/line basis, dealing with harmonics as near as possible to their source.

4.3.2 For economic reasons, it is generally better to correct for harmonic currents on a load area basis as a part of the overall plant power distribution system. This approach requires an engineering evaluation of harmonic current distribution. Acceptable levels of harmonics vary according to total load and power system capacity.

4.3.3 New equipment should be supplied with harmonic filters and magnetic isolation devices to limit steady state total harmonic current to 20 amperes or less at the point of connection of the new equipment.

4.3.4 Harmonic filters should be tuned to a resonant frequency slightly (5-10%) below the most predominant harmonic, which is normally the fifth. If the fifth harmonic is predominant, the filter should be tuned to harmonic order 4.5 to 4.75, or 270-285 hertz.

4.4 Filter Installation

4.4.1 Harmonic filters should not be located directly above heat generating devices.

4.4.2   The harmonic filter should be installed and connected as close as possible to the source of the harmonics. The magnetic isolation device should be connected between the harmonic filter and the system power factor correction capacitors to avoid harmonic overcurrent. The magnetic isolation can be provided by one of two types of devices:

> An isolation transformer with minimum 4% impedance.
> A series reactor with minimum 3% impedance.

4.5   For new equipment installations, the machinery equipment builder should provide the purchaser with spectral harmonic current amplitude information in amperes and in percent of fundamental, through the 23rd harmonic. This is to assist with the integration of the new equipment into the overall plant power distribution system.

4.6   **Filter Application**-See Figure 1 for alternative locations for harmonic filters. The relative merits and drawbacks for each are discussed below.

4.6.1   Generally, power factor and harmonic distortion will be improved, transformer and conductor capacity will be freed up, and current, voltage drop, and I^2R losses will be reduced from the connection point of the harmonic filter in the upstream direction (towards the line side). These benefits will not occur in the downstream (load side) direction.

4.6.2   Generally, harmonic filters not switched with loads can cause overcorrection of power factor during conditions of light loading.

4.6.3   Generally, the further upstream that harmonic filtering is accomplished, the less will be the cost of installation, inspections, service, and maintenance.

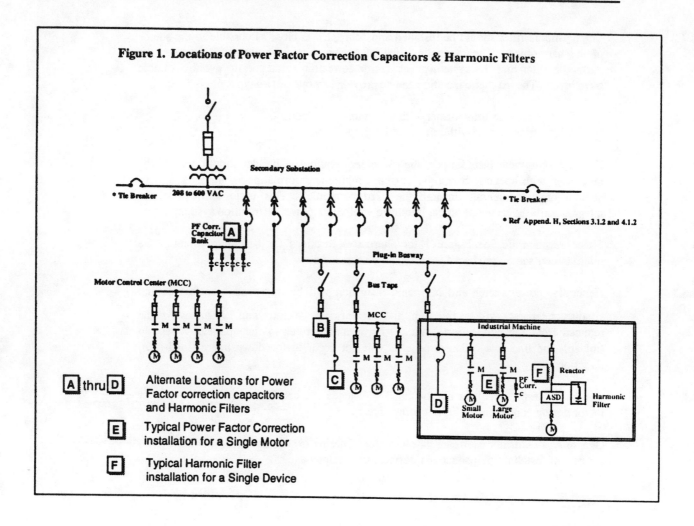

Figure 1. Locations of Power Factor Correction Capacitors & Harmonic Filters

RESPONSE FORM ON SAE HS-1738 USAGE

The objective of this response form is to provide information on:

    (1)   **How SAE Documents are used by industry**

    (2)   **Does this SAE Document meet your needs and expectations**

The SAE Technical Standards Board welcomes your input as a purchaser of this document. We ask that you fill out this form and return it by <u>FAX to (412) 776-0002.</u>

Does your use of this document include:

(1) Procurement of Machine Tools   ___ Yes ___ No
(2) Design of Machine Tools   ___ Yes ___ No
(3) Specification of Machine Tools ___ Yes ___ No
(4)   **Incorporation or reference in Company Standards** ___ Yes ___ No
(5)   **Encourage Reference by other Standards Development Organizations** ___ Yes ___ No
(6)   **Encourage Adoption by other Standards Development Organizations** ___ Yes ___ No

Does this document aid you in reducing costs of your product? ___ Yes ___ No

(1)   **Improve your product** ___ Yes ___ No
(2)   **Provide guidance for new product development** ___ Yes ___ No

Does this document meet your needs? ___ Yes ___ No

Does this document meet your expectations? ___ Yes ___ No

Would you like this information available electronically (CD-ROM)? ___ Yes ___ No

Would you support making this document into an international standard? ___ Yes ___ No

If No's were checked, please provide comments on what changes you recommend.

What other subjects would you like to see developed in this format?
Would you actively support this effort?

I recommend the following changes be considered for inclusion in a future update:

_____

_____

_____

_____

I would like SAE to address the following subject(s) related to Industrial Machinery.

_____

_____

_____

_____

I would be willing to participate in this standards development area.

____ Yes ____ No

NAME _____ DATE _____

ADDRESS _____

PHONE _____ FAX_____

PLEASE FAX TO:
DARLENE CROCKER
SAE
3001 W. BIG BEAVER RD., STE. 320
TROY, MI 48084
(810) 649-0425

SAE HS-1738 1996 EDITION

SAE SUPPLEMENT TO NFPA79
ELECTRICAL STANDARDS FOR INDUSTRIAL MACHINERY 1991 EDITION

PLEASE FAX TO:
DARLENE CROCKER
SAE
3001 W. BIG BEAVER RD., STE. 320
TROY, MI 48084
(810) 649-0425

REQUEST FOR REVISIONS OF SAE HS-1738 1996 EDITION

To insure your request is accepted for inclusion in future revisions of this document, please supply the following information related to your request.

Purpose or Rationale for Request _____

_____

Recommended Changes or Additions
Existing Section(s) _____

_____

_____

Your Specific Change(s) (Please submit marked up pages or new material) to be

included as part of this form _____

_____

Requestor_____**Phone**_____**Fax**_____
Signature _____ **Address** _____
Date _____   _____

COMMITTEE USE ONLY

ACTION TO BE TAKEN _____

_____

SAE HS J1738 CHAIRPERSON_____**DATE**_____